Medical Physics

Volume II
External Senses

Medical Physics

Volume II
External Senses

A. C. Damask

Department of Physics
Queens College of the City University of New York
New York, New York

1981

ACADEMIC PRESS

A Subsidiary of Harcourt Brace Jovanovich, Publishers

New York   London   Toronto   Sydney   San Francisco

ACADEMIC PRESS, INC.
111 Fifth Avenue, New York, New York 10003

United Kingdom Edition published by
ACADEMIC PRESS, INC. (LONDON) LTD.
24/28 Oval Road, London NW1 7DX

Library of Congress Cataloging in Publication Data

Damask, A. C.
    Medical physics.

    Includes bibliographies and index.
    CONTENTS: v. 1. Physiological physics, external
probes.--v. 2. External senses.
    1. Medical physics. 2. Biophysics. [DNLM:
1. Biophysics. QT 34 M491]
R895.D35   612'.014   78-205
ISBN 0-12-201202-X (v. 2)

TO JOHN AND JAY

Contents

Preface

This volume is the second in a set of three which have been designed for premedical students, undergraduates, and postgraduates of all scientific disciplines. Certain constraints are thereby placed both on the topics covered and their level of treatment. A prior knowledge of physiology is not required; the information necessary to understand the physical science is covered in each chapter. The physical science assumes only a prior knowledge of first-year physics and differential and integral calculus. An occasional elementary differential equation is used but its solution has been shown in Volume I, Appendix D. Similarly, the Poisson probability distribution is used, but this also was derived in Volume I, Appendix C. Other than these the reader should encounter no difficulty with the mathematics. In fact, the mathematics is less extensive than in Volume I. This is not due to omission, but rather to the fact that knowledge of the external senses is not yet sophisticated enough to warrant mathematical precision. Not that attempts do not exist in the literature, but the attempts have not succeeded, and some are referenced but not included. Therefore the physics of the sensory processes discussed here is not as directly applicable as was the physics of the processes discussed in Volume I. The physics is contained in the techniques of measurement which have been employed.

It may reasonably be suggested that the senses cannot be treated medically except with mechanical devices such as hearing aids or eye glasses. This is only partly correct because, as will be discussed in this volume, there are other effects on the senses, such as nutritional deficiencies. Additionally, it is intended that this volume be read in conjunction with Volume III, which covers the processing of sensory signals in the synapses and, to some extent, in the brain. When the system is considered in its entirety, it will be evident that medical intervention in the senses need not be confined to the receptor but may involve treatment at other points in the system.

Only the external senses are covered in this volume since far less is known of the internal senses. The electrophysiology and biophysics of the receptors are the primary concern because these are not unique to each species. The pathways to the brain are briefly mentioned for orientation purposes, but the synapsing mechanism and brain electrophysiology are covered in Volume III.

We have witnessed 50 years of electrophysiological mapping of the sensory system and have learned much. It has long been recognized however that little further understanding is possible without new techniques of microchemistry of the neural transmitters. This latter era is just beginning, and descriptions of preliminary techniques and results are given in Volume III, although some of this work is discussed in Chapter 6 on vision. It will be evident from that chapter that there is a great sense of excitement now pervading the field because new discoveries are occurring with increasing frequency. For instance, quite recently the origin of color chromophores in the eye was elucidated, and a discussion of this is included at the end of Chapter 6. This type of fundamental understanding is laying the foundation for future medical treatment of the senses.

The senses are discussed in the order of the increasing amount of research effort expended. As in Volume I, a variety of units has been used, most frequently of the cgs system. The writer's option has been to convert everything to SI units for a self-consistent book or to retain the original investigators' units to simplify the interested reader's understanding of the original papers. The convenience of the readers has been chosen over the elegance of units.

The author wishes to extend his appreciation to Dr. Charles E. Swenberg for critically reading the manuscript and making numerous helpful suggestions.

Contents of Volume I

CHAPTER 1

Introduction

The body has many senses, but we are completely unaware of most of them. For example, continuous monitoring takes place of oxygen and carbon dioxide levels, blood sugar, thermoregulation, digestive juices required, and a host of other functions for which we exert no conscious recognition or control. There are other internal sensors whose signals we do recognize, such as those which tell us the positions of our limbs and the vestibular sense which governs balance. The above senses clearly can be considered as two different types or categories. The first is the senses necessary to the health and well-being of the organism of which we are unaware and generally do not consciously alter. The second is a category of internal senses of which we are aware and to which we respond. A third category includes the senses which respond to external stimuli: taste, smell, touch, hearing, and vision (the classical five senses). This volume considers only these latter senses.

It is tempting to say that these five senses are those of perception by which we perceive the world around us. However, perception is actually a different topic and our treatment of the senses will be considerably more narrow. When an external signal interacts with an organism's receptor, a series of biophysical events occurs within the receptor which generates a signal. The signal becomes part of the information sent to the brain. The brain, in turn, analyzes the signals and dictates the response of the organism, either consciously or reflexively. For example, a shadow moving at a certain optimum speed past a frog's eye will cause a snapping reaction, while a shadow above his eyes will cause him to dive for cover under a rock or into the water (Lettvin *et al.*, 1959; Maturana *et al.*, 1960). Human behavior is generally more analytical than reflexive. Thus, perception might be considered as the analysis by the brain of external stimuli and differs with species. Perception in man is a vast field explored largely by experimental psychologists, and the field is sometimes called psychophysics. This will be discussed briefly in Chapter 7.

We will confine the treatment of the external senses to the biophysical response of the receptors themselves. Such confinement has a distinct advantage. Most animal receptor systems have the same evolutionary history. For example, the chemistry of rhodopsin in the eye of a horseshoe crab is the same as that in a human. Many of the experiments described in this volume have been performed on animals, but correlation with human rece_t tors is always established.

The medical applications of these studies are less precisely defined than those in Volume I, but the research findings reported in this volume are no less important since they lay the foundation for medical treatment of the future. Although much of modern medical treatment is still largely empirical and new discoveries are often serendipitous, it is anticipated that future medical treatment will be based on a firm understanding of the mechanisms involved and biochemicals will be tailored to treat specific deficiencies of the mechanisms. Volume III on synapses and the brain is developed as a continuation of the material in this volume so that a sensory signal and its ultimate mode of conveyance and subsequent reactions can be considered as a continuous process. From this approach, medical treatment of a sense may reasonably involve intervention at any stage of the process from receptor to brain.

When we consider experimental research on the external senses we must stand in awe of some of the great minds of the nineteenth century, such as Mach and Helmholtz, who without modern electronic measuring devices nevertheless used considerable insight in understanding the relation between stimulus and perception, particularly in vision and hearing. Although their research was inspired by still earlier investigators, twentieth century investigators must acknowledge that "If I have seen farther than others it is because I have stood on the shoulders of giants," an expression attributed to Newton but shown by Merton (1965) to have originated with Bernard of Chartres in the twelfth century.

For the past 50 years, active research has been dominated by electrophysiologists. With their fine needle probes, usually hollow glass fibers filled with a conducting fluid, they have been able to probe into cells and observe the electrical response to a stimulus. They have also been able to trace the neural network of animals, including man, from receptor to brain. This is not an easy task because one cannot follow a given nerve; instead a nerve is contacted electrically and the receptor which supplies it with a signal must be located. The nerves of which we speak are actually long extensions, called axons, of neural cells. These were discussed in some detail in Vol. I, Chapter 3. They vary in diameter and in length, with some being about 1 m long, such as that required for a touch signal in a toe to reach the spinal column.

There are so many external stimuli that if they were all transmitted to the brain it would be overloaded with information. It is suspected that to reduce the volume of information axons synapse. That is, several come together at a junction called a synapse, but past the synapse only one axon continues. The transmission across the synapse is chemical in nature rather than the electrical conduction mechanism in the nerve axons. Some external signals go through several synapses before reaching the brain and the amount of information is thereby greatly reduced. A careful examination of the synapse will be found in Volume III.

The axons, or nerves, behave as wires conducting an electrical signal. However, they consist of a thin tubular membrane with a transmembrane resistivity of about 5000 Ω cm. They are filled with a gelatinous substance called axoplasm which has a resistivity of about 50 Ω cm, in contrast to the metal copper whose resistivity is about 10^{-6} Ω cm. Such a conductor constitutes what is called a "leaky" cable, and it was shown in Vol. I, p. 81 that the voltage of a signal in such a cable will fall exponentially.

If the axon diameter is, for example, 0.01 cm, the voltage will fall to about one-tenth of its value in 1 cm. This distance would be perfectly satisfactory in the brain where it is, in fact, a conduction mode. However, in the nerve signals from receptors of external stimuli, the signal needs boosting either continuously or at selected booster stations. This is accomplished by the metabolic energy which has been expended in pumping sodium out of the axon.

The normal ion constituents of body fluids are about 440 mmole/kg H_2O of sodium and 20 mmole/kg H_2O of potassium. However, as explained in Vol. I, in the interior of a resting nerve there is a reversal of concentrations, with 400 mmole/kg H_2O of potassium ions and 50 mmole /kg H_2O of sodium ions. This distribution is established by a "sodium pump" in the interior of the nerve axon. The resting electric potential of the interior of the axon is about 70–80 mV below that of the exterior fluid. When the axon membrane is disturbed physically or electrically, pores open and positive sodium ions rush into the interior. This change in potential effectively "short-circuits" the capacitance of adjacent parts of the axon

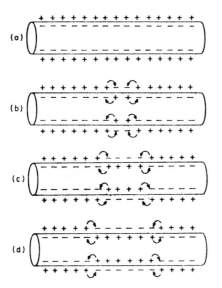

FIG. 1.1   Diagram of passive spreading of the action potential in a nonmyelinated axon. [Redrawn from Guyton (1971).]

membrane and causes their pores to admit sodium ions. This is illustrated schematically in Fig. 1.1 and results in a phenomenon called passive spreading. In this way an impulse of constant potential, called a spike or action potential, is propagated along the nerve axon. After the depolarization of the membrane, which arises from the in-rushing sodium ions, the sodium pump restores the interior negative potential. This requires several milliseconds, during which the axon cannot transmit another spike. In this way signals from sensors travel along the nerve axons as a series of constant potential spikes a few milliseconds apart. Some axons have a myelin sheath surrounding them with exposed parts in regular intervals, about 1 mm apart, called nodes of Ranvier. Such axons can generally conduct signals more rapidly because the depolarization occurs only at the nodes and the signal therefore leaps (saltatory conduction) between nodes. This is illustrated schematically in Fig. 1.2.

Now consider in man a length of myelinated axon running from a finger tip to the spinal column, a length of about 0.8 m. This myelinated nerve would have 800 nodes or boosting stations. In a transmission cable, each boosting station should restore the signal exactly to what it was before the loss as a result of traveling. Only in this way can the quality of the message be maintained. For example (Rushton, 1961), suppose the restoration of the signal was not perfect but only 99% complete at each node. After 800 nodes the signal would be reduced to

$$(0.99)^{800} = 1/3000$$

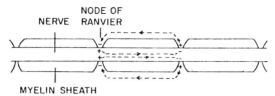

FIG. 1.2   Diagram illustrating the structure of a myelinated axon and the path of saltatory conduction.

of its original size. If, on the other hand, the restoration was 1% greater than the original, the signal would be 3000 times greater than its initial size. Thus, a very small fluctuation from perfect restoration will either overload the system or produce no signal. Clearly, amplitude modulation of the signal is not possible. The only perfect type of signal is one that is on or off, that is, a unit change as shown in Fig. 1.3a. But there is no more information in a signal of type 1.3a than in its differential, type 1.3b. But 1.3b alternates regularly and no information is contained in the change of sign. So a series of positive pulses, Fig. 1.3c, carries the same information as Fig. 1.3a, but with far less energy expended. Thus, from general considerations, a signal of type 1.3c is the preferred one and is the type which the nerve carries.

A single pulse on an extended time scale is shown in Fig. 1.4. The small early pulse is the experimenter's initiating voltage. The shape of the pulse is not important, only its presence or absence matters. The possibility that two or more pulses are too close to be discriminated is precluded by the refractory period of about 3 msec. So a single human nerve can transmit about 300 impulses/sec, or in a typical reaction time of 0.1 sec about 30 *bits* of information, i.e., impulse present or absent, can be transmitted. The number of possible arrangements of 30 bits is 2^{30}. This is about 1 billion

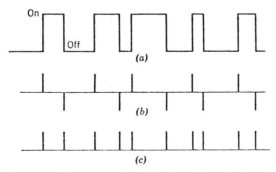

FIG. 1.3   Possible types of all-or-none signal. [Reprinted from Rushton (1961) by permission of the MIT Press. Copyright 1961 by the Massachusetts Institute of Technology.]

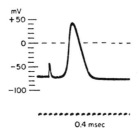

FIG. 1.4 Action potential of a non-myelinated squid axon. The vertical scale is the internal potential in millivolts. [Unpublished work by A. L. Hodgkin and R. D. Keynes quoted by Hodgkin (1958).]

different possible messages sent to the brain. As mentioned earlier, this is far more information than the brain can handle, so the amount is considerably reduced by synapsing.

In every chapter in this volume recordings of these spikes or action potentials will be illustrated. This is not redundancy. Rather, there is a code from each type of receptor which tells the brain a variety of things, such as the location of the receptor, the strength of the stimulus, the rate of change of the stimulus, and the intensity relative to adjacent receptors. Although it is a code which has not yet been broken, the efforts of some investigators in this direction will be indicated. It should be noted, however, that not every stimulus has sufficient intensity to generate an action potential. Usually the biological receptor potential must be altered by 20–30 mV before an action potential is propagated along the afferent neuron. This disturbance of the receptor potential is called the generator potential and may be too weak to cause the creation of an action potential.

Conduction processes in either myelinated or nonmyelinated axons will also be discussed. A number of investigators have been concerned with the reason for both types in the body. Their studies, as well as a theoretical model, have been considered by Rushton (1951). The argument will be given briefly. It was shown in Vol. I, p. 83, that, on the basis of the Hodgkin–Huxley model of nerve conduction, the velocity of transmission is proportional to the square root of the diameter of the axon in nonmyelinated nerves. Rushton reviews the evidence that in myelinated nerves the velocity of transmission is proportional to the diameter of the axon. For the survival of an organism certain information must be transmitted as rapidly as possible, but because of the delay in signal processing through the synapses it is sometimes advantageous to enrich the information by means of a group of smaller, slower transmitting nerves, instead of a single large one. Thus, organisms have a variety of sizes of both types of axons, and the size and type associated with a sense is the evolutionary answer to the importance of the signal. In Fig. 1.5 a plot of the velocity of myelinated (linear) and nonmyelinated (quadratic) versus diameter is shown, based on data obtained from vertebrates. It is seen that although the saltatory

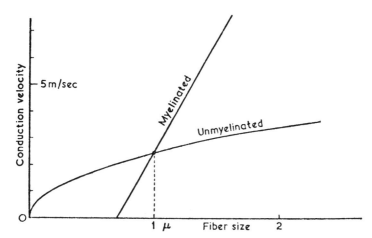

FIG. 1.5 Conduction velocity versus fiber diameter for myelinated and nonmyelinated nerve axons. [From Rushton (1951).]

conduction mechanism of the myelinated fibers is more rapid at large diameters, it becomes less rapid than the nonmyelinated fibers at small diameters. Thus, from the standpoint of energy efficiency of an organism, higher velocity of signal transmission below 1 μm diameter can be achieved without expending metabolic energy to create and maintain myelinated fibers. Not surprisingly, 1 μm is the critical diameter above which all fibers are myelinated. However, the above considerations are only for conduction velocity. Deviations may be expected when some other function is more important, such as an above normal speed of recovery from depolarization.

REFERENCES

Guyton, A. (1971). "A Textbook of Medical Physiology." Saunders, Philadelphia, Pennsylvania.

Hodgkin, A. L. (1958). Ionic movements and electrical activity in giant nerve fibers, *Proc. R. Soc. London Ser. B* **148**, 1.

Lettvin, J. Y., Maturana, H. R., McCulloch, W. S., and Pitts, W. H. (1959). What the frog's eye tells the frog's brain, *Proc. Inst. Radio. Eng.* **47**, 1940.

Maturana, H. R., Lettvin, J. Y., Pitts, W. H. and McCulloch, W. S. (1960). Physiology and anatomy of vision in the frog, *J. Gen. Physiol.* **43**, Suppl. 129.

Merton, R. K. (1965). "On the Shoulders of Giants." Harcourt, New York.

Rushton, W. A. H. (1961). Peripheral coding in the nervous system, *in* "Sensory Communication" (W. A. Rosenblith, ed.). MIT Press, Cambridge, Massachusetts.

Rushton, W. A. H. (1951). A theory of the effects of fibre size in medullated nerve, *J. Physiol.* **115**, 101.

CHAPTER 2

Gustation

INTRODUCTION

Chemical sensibility is very primitive. Protozoa will take an avoiding action when chemical irritants are placed in the water they inhabit, as will jellyfish and all higher forms of animal life. As more complex forms of life developed the chemical detectors became more localized and, among the vertebrates, the fishes and amphibians have chemical receptors distributed over large areas of the skin. When animals left the sea the skin became hardened and lost its chemical receptors. Because there was survival value in identifying separately airborne molecules and those to be ingested through the mouth, the chemical senses separated into olfaction and gustation (taste). A part of the brain, the olfactory bulb, developed to analyze and identify the signals from the chemical receptor cells in the nose. Groups of receptors, called taste buds, which respond to four specific tastes formed on the tongue, but their signal does not go to the olfactory bulb. We will consider taste in this chapter.

The sense of taste in adult humans is of four basic types: sweet, sour, salty, and bitter, although, as will be shown, these categories are not precise. The stimulus originates from a chemical contact with a receptor cell on the tongue. This generates an electrical signal in a nerve fiber in the tongue which connects through a synapse with one of a large bundle of nerve fibers, called the *chorda tympani*, then through a rather complicated route to the brain.

The actual tasting of a substance such as a food involves olfaction as well, and the experiments must therefore be carried out with care. Most, but not all, of the electrical signals in the nerves are measured on laboratory animals, while the subjective experiments which require opinion and judgment are performed on humans. A comprehensive review of the literature is given by Sato (1980).

TASTE RECEPTORS AND NERVES

The tongue of an adult human is covered with little projections which make it rough. These are called *filiform papillae* and are not involved in taste but serve to grasp and maneuver food in the mouth. In some animals they are so rough that they can lick bones clean of flesh. Among these papillae are another kind called *fungiform papillae* which contain the taste buds. There are about 9000 of these on the upper surface and sides of the tongue of an adult human. On the upper surface of the tongue near the back are seven to ten large *circumvallate papillae* which also contain taste buds. Although all taste buds respond more or less to all four tastes, the dominant characteristics appear to be grouped at certain areas of the tongue, Fig. 2.1. The tip of the tongue responds primarily to sweet, the sides of the tip to salt, the sides of the center to sour, and the upper rear surface to bitter.

The cells within a taste bud were classified by early histologists as *sustentacular* (supporting) and *gustatory*, although modern electron microscopists consider this differentiation oversimplified. Figure 2.2a shows drawings of two types of taste buds on a human tongue, with associated nerve fibers, and Figure 2.2b shows electron micrograph of a taste bud of a

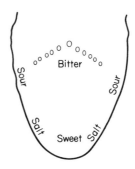

FIG. 2.1 The areas of the tongue where the four tastes are most easily sensed. In these areas the taste buds are very numerous. All four tastes are also sensed, but less readily, over all areas.

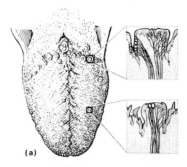

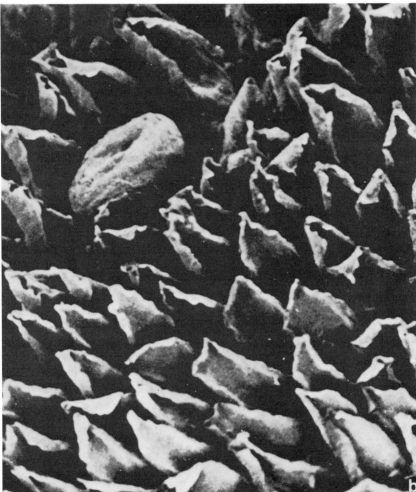

FIG. 2.2 (a) A human tongue illustrating the fungiform papilla, lower right insert, and circumvallate papilla, upper right insert. The inserts show the location (open circles) of the taste buds on each and the associated nerve fibers. [From Henkin (1970).] (b) Surface of a rabbit tongue shows a single fungiform papilla with the surrounding filiform papillae that resemble leaves. The fungiform papilla contains the taste buds. [From Beidler (1970). Both (a) and (b) are from articles in J. F. Bosma, ed., "Oral Perception and Sensation," 1970. Courtesy of Charles C. Thomas, Publisher, Springfield, Illinois.]

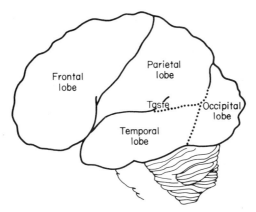

FIG. 2.3   Schematic of the brain with the location of taste (side view).

rabbit. It has been learned that each taste cell has a double membrane, each 75 Å thick and about 100 Å apart. At the exposed end of the membrane there are fingerlike projections called *gustatory microvilli*. These projections are about 2 μm long and 0.1 μm wide and make contact with the aqueous solution (saliva) which covers the tongue. These structures are believed to be the chemically excitable regions. The gustatory cells have a short lifetime, about seven days, and are constantly being replaced.

There are several gustatory cells within each bud and these are connected through a synaptic terminal to a nerve fiber. The nerve fibers are collected into a small bundle of fibers called the chorda tympani which, in turn, leads into the seventh cranial nerve called the facial nerve. It should be noted that not all of the taste buds in humans follow this route; those at the back of the tongue join a different nerve as do a few from the base of the tongue. However, all of the electrical impulses go to the *thalamus*, which is at the middle of the brain, and from there they are transmitted to a part of the brain in the limbic lobe called the *parietal opercular-insular* area. The general location of this region is seen in Fig. 2.3. It is not located on the outer surface, but somewhat to the interior.

THE FOUR TASTES

Most subjective experiments with humans have been performed on the four basic tastes, sweet, sour, bitter, and salt, but it is believed by many investigators that alkaline and metallic tastes should be included in the basic list. Many things taste like some proportionate mixture of the above basics, which makes experiments difficult. Olfaction must be avoided and temperature must be body temperature. Under these circumstances, a blindfolded subject with a nose clamp cannot distinguish between grated onion, apple, and turnip—they all taste slightly sweet. Furthermore, just as one odor can mask another, so can tastes.

The sour tastes we commonly encounter are acetic (vinegar) and citric (lemon) acids. However, not only do these organic substances have tastes other than simply acid, they also are only partially dissociated ionically. If, as some have postulated, it is the hydrogen ion concentration which causes the sour taste, then measurements should be made on a fully dissociated acid such as hydrochloric. However, experiments have shown that hydrochloric acid can just be tasted at $N/800$ (1/800 of a normal solution) while acetic acid can just be tasted at $N/200$, although pH measurements indicate that the HCl solution has a concentration of free hydrogen ions four to five times higher than that of the acetic acid solution. Many other experiments were performed with anion changes, buffered solutions, etc., with the general conclusion that both the concentration of the anion and its ability to penetrate the tissues have effects on the ability of the H^+ ion to stimulate a sour taste.

The salty taste is not confined to NaCl, the chlorides of potassium, ammonium, and calcium all taste similar. It is tempting to conclude that it is the Cl^- ion rather than the Na^+ which causes the salty taste. However, NaBr and NaI also taste salty. Kionka and Strätz (1922) compared 18 salts and divided them into the following three groups:

(1) predominantly salty: NaCl, KCl, NH_4Cl, LiCl, RbCl, NaBr, NH_4Br, LiBr, NaI, LiI;
(2) both salty and bitter: KBr, NH_4I;
(3) predominately bitter: CsCl, RbBr, CsBr, KI, RbI, CsI.

TABLE 2.1[a]

Molar concentration	Taste of NaCl	Taste of KCl
0.009	No taste	Sweet
0.010	Slight sweet	Sweeter
0.015	Sweeter	Still sweeter
0.020	Sweet	Sweet, bitter
0.030	Strong sweet	Bitter
0.040	Salty sweet	Bitter
0.050	Salty	Bitter, salty
0.070	Saltier	Bitter, salty
0.100	Still saltier	Bitter, salty
0.200	Pure salty	Bitter, salty, sour
0.500	Pure salty	Bitter, salty, sour

[a] From von Skramlik (1926).

They considered that the taste of salt depends on the cations Na^+, K^+, NH_4^+, Li^+, Rb^+, and Cs^+, all of which give intensity, and the anions Cl^-, Br^-, and I^-, which give the character. Other investigators have come to somewhat different conclusions (Moncrieff, 1967).

von Skramlik (1926) has shown that the taste sensation may vary with the concentration. As seen in Table 2.1, NaCl and KCl taste sweet at low concentrations, while KCl tastes bitter at high concentrations. Thus, the fundamental basis for salt taste is not well established.

The sweet taste is no less confusing. Sugars contain a group –CO–CHOH– which, if combined with at least one hydrogen atom, become what is called a glucophore. These, however, have no chemical relation to saccharin, which is very sweet.

It is apparent from the above structures that if saccharin is modified slightly it becomes tasteless. Although much research has gone into this area (see Moncrieff, 1967), very little is understood.

The bitter taste, like the sweet, occurs in many disparate organic and inorganic compounds. For example, magnesium and ammonium salts are often bitter, as are quinine and strychnine.

The thresholds for taste are given as follows in concentrations just detectable:

sweet	0.7% sugar, 0.001% saccharin
salt	0.055% NaCl
sour	0.0045% HCl
bitter	0.00006% strychnine hydrochloride

This list shows increasing sensitivity downward, if natural sugar is used. Thus, the body has extreme sensitivity to bitter taste, which must have had a high survival value in evolution.

ELECTRICAL SIGNALS IN THE TASTE BUDS

As mentioned above, a single taste receptor seems to be responsive to all four tastes, although with different magnitudes of response. With the introduction of microelectrode techniques, Kimura and Beidler (1961) inserted an electrode into a taste bud of the tongue of a rat. With repeated trials, a location could be found which had a lowered potential, and the

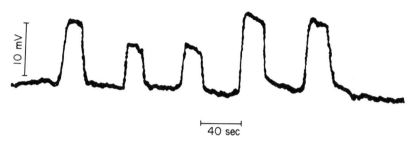

FIG. 2.4   Receptor potential of a rat taste cell in response to $0.1\,M$ of NaCl, KCl, NH$_4$Cl, CaCl$_2$, and MgCl$_2$ applied to the tongue surface with water rinses between stimuli. [From Kimura and Beidler (1961).]

electrode was assumed to be within a single taste cell. The tongue was then wetted with the stimulating solution and the resulting potential changes observed.

Example microelectrode recordings from a single taste cell of a rat are shown in Fig. 2.4. The separate peaks are for $0.1\,M$ solutions of the various salts listed in the captions, with the tongue being washed with water between applications. This figure illustrates the similarity of electrical response of a single taste cell, although the magnitudes are different. Figure 2.5 shows the response of a single taste cell of a hamster to the four tastes in order: salt, bitter, sweet, and acid. The molar concentrations of these are different, but this is not a quantitative measurement because it is not known how much reaches the taste cell. This figure demonstrates that an individual taste cell is, or may be, responsive to all four tastes, although the intensity of response to each type may vary. No action potentials of electrical signals have been recorded from a single taste cell since they would be much larger than the 10 mV potentials shown in Fig. 2.4. The illustrated signals are probably generator potentials and the addition of several from different cells will generate an action potential.

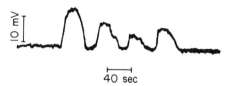

FIG. 2.5   Receptor potential of a hamster taste receptor in response to $0.1\,M$ NaCl, $0.02\,M$ quinine hydrochloride, $0.5\,M$ sucrose, and $0.01\,M$ HCl applied to tongue surface with water rinses in between. [From Kimura and Beidler (1961).]

ELECTRICAL SIGNALS IN NERVE FIBERS

In further experiments the chorda tympani was dissected at the base of the tongue and both single and groups of nerve fibers were separated from the bundle. A single nerve fiber, however, probably carries the signal from several gustatory cells within a bud, as suggested by the data of Table 2.2.

TABLE 2.2

	Estimated number of taste buds[a]	Myelinated fibers in the chorda tympani[b]
Sheep	10, 000	3400
Pig	15, 000	4400
Goat	15, 000	3300

[a] From Moncrieff (1967).
[b] From Kitchell (1963).

Measurements on a single fiber reveal the action potentials. Figure 2.6 shows such action potentials in a single fiber when a rat's tongue is bathed in solutions of NH_4Cl with increasing molarity. The spike (action potential) height is 100 μV and the time scale is 0.2 sec per interval, marked at the bottom. It is seen that saturation of the signal pattern occurs at about the third concentration, $0.25 M$. Such saturation of the signal is best measured by an integrated electrical response of a bundle of fibers. A typical measurement of the increase in intensity of the integrated electrical signal, with increasing concentration of NaCl solution, is shown in Fig. 2.7. A number of such measurements resulted in the curve of the integrated electrical response, with increasing molar concentration of NaCl, shown in Fig. 2.8, in which the saturation of electrical activity is evident. The saturation level is different for different salts, Fig. 2.9, which is somewhat surprising because one expects the electrical activity to be limited by the conduction properties of the nerve rather than some property of the stimulus.

A fatigue effect, which probably occurs in the synapse rather than the nerve, has also been observed. The effect is illustrated in Fig. 2.10, in which the number of impulses per second is plotted against time for different salts. It is seen that there is a rapid decrease in a fraction of a second with a low steady-state value remaining. This fatigue effect is a common response of the senses.

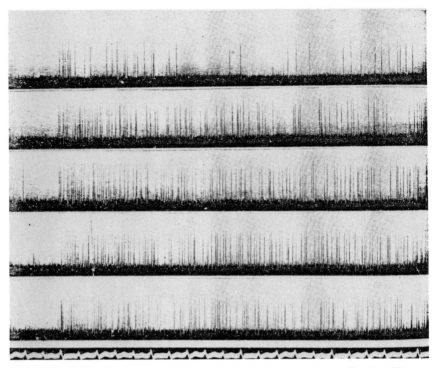

FIG. 2.6 Electrical activity recorded on a single chorda tympani nerve fiber for different concentrations of NH_4Cl solution applied to the tongue of a rat. Concentrations applied from top to bottom: $0.01M$, $0.05M$, $0.25M$, $0.5M$, and $0.75M$. Spike height 100 μV, time scale one interval on bottom line in 0.2 sec. Note saturation of activity by $0.25M$ and compare with integrated saturation of Fig. 2.8. [From Beidler (1953).]

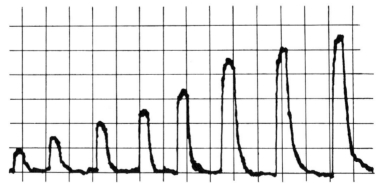

FIG. 2.7 Integrated electrical responses of chorda tympani to NaCl solutions of different concentrations applied to a rat's tongue interspersed with water rinses. Concentrations from left to right: $0.005M$, $0.01M$, $0.025M$, $0.05M$, $0.10M$, $0.25M$, $0.5M$, and $1M$. [From Beidler (1953).]

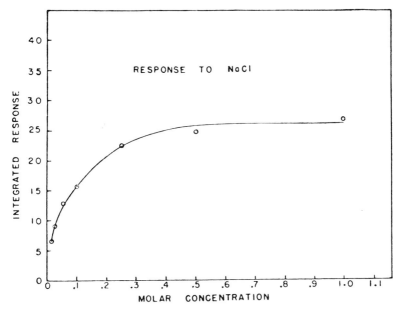

FIG. 2.8 Curve of integrated response recorded on chorda tympani of a rat for various concentrations of NaCl solutions flowed over the tongue. [From Beidler (1953).]

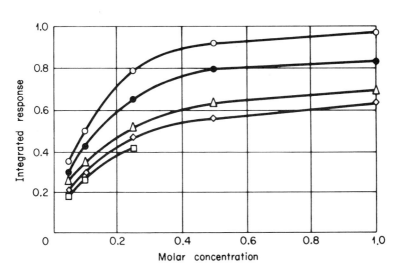

FIG. 2.9 Curves of the integrated response recorded on chorda tympani of a rat as a function of concentration of different salts. Top to bottom: NaCl, Na formate, Na acetate, Na propionate, Na butyrate. [From Beidler (1962).]

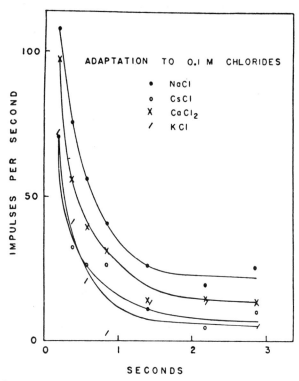

FIG. 2.10   Response of a small group of fibers in the chorda tympani to various 0.1 M
chloride solutions flowed over tongue of a rat. [From Beidler (1953).]

HUMAN PSYCHOPHYSICAL RESPONSE

The chorda tympani is part of the facial nerve that passes behind the
eardrum (see p. 147), and often some of the fibers are found within the
layers of the eardrum itself. Otosclerosis, which produces deafness, is
caused by sclerotic changes in the auricular bones behind the eardrum (see
Chapter 5). An operation is possible to correct this condition, but a
temporary removal of the eardrum is required. The chorda tympani is
immediately encountered and it must be stretched or, in some cases,
severed. Bull (1965) has reported that a significant number of patients
whose chorda tympani had been stretched, or partially or wholly severed,
find an abnormal taste sensation afterward, usually describing the tastes as
metallic or bad. Pity the deaf gourmet who may have to decide which sense
he prefers to have!

The exposed chorda tympani on conscious patients has permitted a
combination of the measurement of electrical signals and subjective taste.

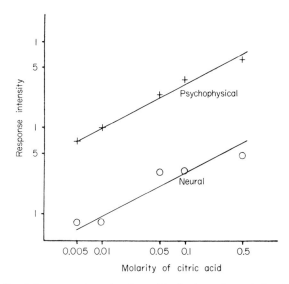

FIG. 2.11   Graph for one patient showing subjective intensity and electrical response of the chorda tympani plotted against molarity of citric acid on the tongue, log–log scale. [From Borg *et al.* (1967).]

Figure 2.11 shows a log–log plot of the subjective change in taste sensation to changing solution concentration of citric acid on the tongue and a similar plot of the electrical response of the nerve. Note that the curves have the same slope but that the curves have not been shifted, i.e., normalized, because there is no standard of comparison of taste sensation with electrical signals. The comparison of psychophysical measurements will be considered in more detail in Chapter 7.

CHEMICAL EQUILIBRIUM MODEL OF TASTE INTENSITY

Figure 2.10 shows that the taste response is rapid, which suggests that the stimulant does not have to diffuse into the gustatory cell but rather acts upon touching its surface. This is further indicated by the rapidity with which a water rinse stops the activity. If the contact of the stimulant with the cell initiates an electrical impulse, the energy must be supplied by the contact, and it logically follows that there must be a free energy change upon contact. It is not necessary to consider the details of the mechanism in order to apply chemical equilibrium theory, as Beidler (1954) has done.

Consider a stimulus of concentration C and initially N receptor sites. The C's interact with the N with a rate constant K_1 to form the bonded stimulus–receptor complex, Z. Since the bond between stimulus and receptor is not strong or permanent, there is a back reaction which separates them with a rate constant K_2. Note that when Z is formed there are no longer N receptor sites for further C's to bond with, instead there are $N - Z$. We may write the chemical reaction as

$$\text{stimulus} + \text{receptor sites} \underset{K_1}{\overset{K_2}{\rightleftharpoons}} \text{stimulus–receptor}$$

$$C + N - Z \underset{K_1}{\overset{K_2}{\rightleftharpoons}} Z \tag{2.1}$$

The rate of change of Z with time (see Vol. I, Appendix A, Section A.7) is written

$$dZ/dt = K_1 C(N - Z) - K_2 Z \tag{2.2}$$

When steady state is reached, i.e., there are as many back reactions as forward ones, Z no longer changes with time and therefore

$$dZ/dt = 0$$

With this condition, Eq. (2.2) may be written

$$K_1/K_2 = Z/C(N - Z)$$

and the ratio of forward to back reaction is defined as K, the equilibrium constant, $K_1/K_2 = K$, thus

$$K = Z/C(N - Z) \tag{2.3}$$

One may now assume that the magnitude of the neural response R is proportional to the number of stimulus–receptor complexes formed, $R = aZ$, where a is a proportionality constant. Also, the maximum response R_s occurs at a high concentration when all the receptors are bonded to C's so that $R_s = aN$. Substituting for Z and N in Eq. (2.3) gives

$$K = R/C(R_s - R)$$

Inverting, removing the parentheses, and rearranging yields the equation

$$C/R = (C/R_s) + (1/KR_s) \tag{2.4}$$

which is called the fundamental taste equation. It is not a profound or unique equation and occurs frequently in many fields of science, e.g., in surface science it is analogous to the Langmuir absorption isotherm (pp. 62–63). It is very satisfying that taste intensity has been shown by Beidler to obey this simple relation which also is used in other receptor binding studies. It will be shown in Chapter 3 that olfaction intensity also obeys the same relation.

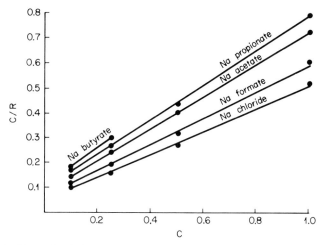

FIG. 2.12 The ratio of the molar concentration of the stimulus to the magnitude of the neural response plotted against the molar concentration of the stimulus for the data of Fig. 2.9. [From Beidler (1954).]

If quantitative data obey Eq. (2.4), a plot of C/R versus C will yield a straight line with a slope of $1/R_s$ and $1/KR_s$ as the intercept. The only unknown in the equation is K, which can be evaluated from the intercept. When the data of Fig. 2.9 are plotted in this manner they yield the straight lines shown in Fig. 2.12.

The free energy ΔG^0 of the reaction may be computed from the equilibrium constant K by the expression (see Eq. (A.34'), Vol. I)

$$\Delta G^0 = -RT \ln K$$

where R is the gas constant and T the absolute temperature. The constants and free energies have been computed from Fig. 2.12 by Beidler and are given in Table 2.3. Note that the values of the change in free energy decrease slightly with increasing size of the anion.

TABLE 2.3

Salt	R_s	K	ΔG^0 (cal/mole)
Sodium chloride	2.17	9.80	− 1.37
Sodium formate	1.89	9.00	− 1.32
Sodium acetate	1.55	8.55	− 1.29
Sodium propionate	1.44	7.58	− 1.22
Sodium butyrate	1.30	7.72	− 1.23

If no work is done on or by a system, the Gibbs free energy is the same as the Helmholtz free energy and they are given by the relation [Eq. (A.4), Vol. I]

$$\Delta G = \Delta H - T \Delta S \qquad (2.5)$$

The physical meaning of the terms is the following. Chemical bonds in a reaction may be broken and others formed; ΔH is the sum of the energy difference between the bonds formed and those broken in a particular reaction. The term ΔS is the entropy change which occurs when the molecules are rearranged configurationally or mixed together, and entropy changes also arise from the changes in vibrational spectra associated with these rearrangements.

Beidler has made the following argument that the association of a stimulus ion or molecule with a receptor cell is a physical one rather than one that involves chemical bonding. The free energy of the reaction given in Table 2.3 is very low compared with that of any chemical reaction which involves the breaking and reforming of bonds. According to Eq. (2.5) such a low free energy could have two origins: (1) either no chemical bond

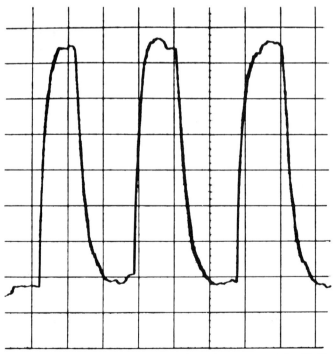

FIG. 2.13 Neural response of chemoreceptors of a rat's tongue to 0.5M NaCl at 25°C, 30°C, and 20°C. [From Beidler (1954).]

changes occur, in which case $\Delta H = 0$, and if ΔS is small, ΔG would be correspondingly small or (2) bonds are broken and reformed and ΔH is large, but $T\Delta S$ is also large and their difference small. Beidler was able to decide between these alternatives by measuring the response of the electrical signal when a rat's tongue was bathed with the same concentration solution at different temperatures. Figure 2.13 shows that there is no measurable temperature dependence. Therefore, the $T\Delta S$ term is small and ΔH itself must be small. There are no bond breakages which have small energies, and it is therefore reasonable to conclude that since ΔH is small it must actually be zero and no bonds are broken in the reaction. Thus, the attachment of the stimulus to the receptor is a physical one which involves small changes in entropy. The energy supplied by the attachment is apparently sufficient in a local region to change the membrane potential by the amount required to create an action potential.

THE TASTE FOR SALT

Sodium is a necessary ion in the plasma for physiological performance. For example, its role in nerve signals was discussed in Vol. I, Chapter 3. Therefore, from an evolutionary viewpoint, it is reasonable that a taste receptor for salt should exist and that a preference for this taste should be moderated in some way by the sodium level in the blood plasma. It is believed that the salt level in the blood is sampled as it flows through the adrenal cortex and a preference urge originates there. The reason for this belief is that animals who have had their adrenal cortex removed have an unsatisfiable urge to continually ingest salt.

The existence of the preference for salt is also confirmed by animal studies. Grazing ruminants often ingest a higher potassium than sodium content in certain seasons and are then attracted to salt water or salt licks. They will ingest a certain quantity and then apparently become satisfied. Controlled laboratory experiments on rats have also demonstrated this phenomenon. If two water sources are available to a rat, one salty and one not, he will prefer the plain water. When he has been fed a reduced sodium diet he will prefer the salt water. The feedback signal from the adrenal cortex is not prompt, however, because once he has been drinking the salt water he will continue to prefer it for many days after his sodium requirements have been reestablished.

There is a distinct difference between rats and humans in salt taste, and the interpretation of human behavior on the basis of animal experiments must be made with caution. Figure 2.14 illustrates this difficulty. The electrical response in the chorda tympani of three human subjects and one rat is shown as a function of time. It is seen that after an initial fast rise

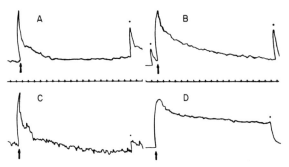

FIG. 2.14 Integrated chorda tympani response to continuous flow of 0.2 M NaCl over the tongue. A, B, and C are for humans and D is a rat response. Arrows indicate time zero and dots on baseline indicate 10-sec intervals. [From Diamant *et al.* (1965).]

there is a slow decrease in the signal intensity, which may take a minute or longer. The important observation in this experiment is that the human intensity returns to zero while the salt is still on the receptor, but the signal for the rat attains a constant positive level. The human becomes more sensitized to salt taste than the rat.

In order to taste salt, it obviously must be in a concentration greater than that of the concentration in the saliva. In another study the subjective threshold of salt taste was measured on human tongues. The tongue of an individual was held under a rinsing stream of a given molar concentration, called the adapting solution, and plotted on the abscissa of Fig. 2.15. The tongue, when rinsed, was then quickly placed in another stream and the subject would state if he could taste salt or not. With different concentrations of salt in a variety of second streams a threshold for salt taste could be determined. The diagonal line is the locus of points of equal concentration of adapting solution and taste solution; obviously all thresholds must lie above this line. The open circles are the thresholds for three subjects plotted against their salivary molarity as the abscissa. These results clearly indicate that the threshold is significantly altered by adaptation to water or weak saline solutions. Furthermore, the threshold values obtained when the tongue is adapted to saliva closely approximates the value to be expected if salivary NaCl were the primary factor influencing NaCl thresholds.

It is indicated by these experiments that in humans the desire for the taste of salt can exist above the body's requirement for salt. The combined effects of (1) the tasting of it which necessitates a concentration above that of the saliva, (2) the losing of the taste of it after a minute or so, at least within reasonable concentrations, and (3) the adapting to a new concentration preference which can last for many days (at least in rats) results in a variety of steady-state sodium intake and excretion quantities among

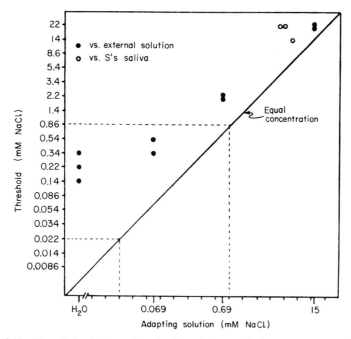

FIG. 2.15 The relation between the adapting solution on the human tongue and the taste threshold for NaCl. Open dots indicate molarity of saliva. [From McBurney and Pfaffman (1963). Copyright 1963 by the American Psychological Association. Reprinted by permission.]

humans. In simpler terms, some like more salt on their food than others. Is high salt intake deleterious? Dahl (1958, 1960, 1967; Dahl *et al.*, 1970) and others have accumulated data which show that countries in which the inhabitants have a high salt intake in their food have a high incidence of hypertension (high blood pressure), whereas the converse is true for those countries with low salt intake. Although a cause and effect relationship is not proven, circumstantial evidence has been accumulated which indicates that a low salt diet is probably healthier for the general population, but not necessarily for a given individual.

Dahl and his associates carried out an interesting series of experiments on rats. Upon feeding them a high salt diet, they found that some rats suffered hypertension while others did not. They then separately bred the two groups of rats for three generations, those who became hypertensive in one group and those that did not in the other. Upon feeding the high salt diet to the sensitive group, all suffered high blood pressure and most died during the experiment, while the other group was unaffected. On the basis of these experiments they hypothesized that there is some genetic metabolic disorder which salt can aggravate. Because a reduced salt diet is an effective treatment in human hypertension, they suggested that there is a similar genetic difficulty in humans, and in a statistically distributed population lower salt intake is probably healthier.

Salt seems to be an acquired taste. Reports by anthropologists indicate that primitive tribes do not like the taste of salt at first, but they later acquire a taste preference for it. In another experiment Dahl *et al.* (1970) fed commercially prepared baby food to his salt sensitive rats and many died of hypertension. It seems that the baby food manufacturers flavor their product to please the mother, who usually tastes it before feeding her baby. The mothers, having a high preference for salt, would thus, by their taste selection of the food, instill the preference in their babies and thereby perpetuate the higher salt intake in the United States. How much salt is enough? Dahl pointed out that the low fraction in mother's milk should be adequate since millions of years of evolutionary development produced the correct dietary balance for human children. One baby food manufacturer now advertises that no salt is added to his product. The final decision will undoubtedly be made in the marketplace, and a rejection of the reduced salt product by an uneducated public will be to their ultimate disadvantage. In a country where children thrive on pretzels and potato chips, hypertension may become preferable to salt deprivation.

TASTE AND NUTRITION

The role of taste in nutrition is extremely complex and there are no easy formulae. Experiments are difficult because, while taste is a subjective sense, hunger is an extremely strong motivating force through which taste aversions and psychological barriers are generally suppressed. It is also possibly premature to discuss nutrition before Chapter 3 which deals with olfaction, because both senses are used in eating habits, but nevertheless some experiments can be described. Children have extra taste buds on the whole upper surface of their tongues and on the insides of their cheeks. These are possibly the origin of their partiality for sweets, which probably had high survival value in the evolutionary scale, considering the rate at which children consume energy. Sweetness, however, is purely a taste sensation. As a human grows older he (hopefully) loses these extra taste buds and the other three tastes dominate. Coupled with these, odor begins to play a role as adults develop a preference for partly decomposed foods such as cheeses, usually the "smelly" kinds, which children dislike. The subjective aspect is also difficult to apply to humans from animal studies. Not only have variations in preferences for different sugars been found in animal species, but false sweets such as saccharin apparently taste pleasant to rats but bitter to dogs, while some humans also dislike its taste. Studies of genetic differences in tastes have been made. In spite of these seemingly overwhelming handicaps, excellent experiments have been performed which begin to give us some understanding of the role of taste in nutrition.

There is unquestionably an instinct at work in the nutrition of lower animals. Experiments describing the salt preference to replace sodium difficiency were discussed in the preceding section. It is known that in certain parts of Africa where the soil is poor in phosphates, cows search for and eat old bones. In experiments by Pyke (1944), rats were kept in cages containing cups of purified carbohydrate, protein, fat, and about a dozen other minerals and vitamins. Although these substances are never encountered naturally in pure form, the rats chose a diet which was nutritionally perfect and made more economical growth than the diet which was usually compounded for them. Although it has often been suggested that humans would do the same, the evidence for this is still controversial.

Taste and smell in animals seem to have little or no role in the long term quantitative regulation of food and water intake, as experiments by Teitelbaum and Epstein (1962, 1963) and others have shown. If the nutrient content of solid powdered feed for rats is decreased by mixing it with nonnutritive cellulose, the animal will eat enough to survive with nonnutritive amounts up to 75% by weight. If water is added to a liquid diet, reducing the nutrient content to as little as 2%, the animal will continue to consume an adequate amount of nutrient even though overhydration results. Direct manipulation of taste seems to have no effect. The food can be sweetened with dextrose or made bitter with quinine hydrochloride and the rats continue to maintain normal caloric intake. For water intake the situation is quite similar. When quinine hydrochloride is added, the rats continue to drink it up to a concentration of 1%, which is the lethal quantity for normal water intake. The animals continue to drink it until they actually become poisoned.

In order to completely remove taste and smell from the test, a feeding tube was passed through a rat's nostril directly into his stomach. The rat was taught to inject liquid food by pressing a pedal. A small amount was injected with each depression of the pedal and the food was made extremely bitter so that regurgitation was not a method of tasting. The rats suffered no transition difficulties in going from oral to tube (intragastric) feeding. They maintained normal nutritional intake even when the load per pedal depression was changed. When the nutrient content of the liquid was changed by 50% in either direction, the rats promptly adjusted their feeding habits. The absence of regurgitated tasting is indicated in Table 2.4, which shows the average quantity of intake over a several day period in the first column, followed by the daily consumption after the food was made bitter.

It is seen that the rats who can taste the bitter quinine take about three days to adjust to the adulteration, while the tube-fed rats require no adjustment time. These experiments show that in rats the central neural mechanism that controls food intake can operate effectively without sensory information, such as taste, smell, or mouth feel, and without feedback from the muscles that are involved in chewing and swallowing.

TABLE 2.4[a]

	Pure diet	Quinine adulterated diet		
		Day 1	Day 2	Day 3
Oral feeding	42 ml	13	24	36
Intragastric feeding	29	33	33	30

[a] From Epstein and Teitelbaum (1962).

EXPERIMENTS ON BRAIN-DAMAGED ANIMALS

The above studies were made on normal animals, and from these it may be concluded that there is such a powerful urge to eat and drink that all other barriers such as taste are overcome. But regulation cannot exist without motivation, and in brain-damaged animals the existence of a motivating force such as taste can mean the difference between life and death. One of the motivation sources for feeding arises in the hypothalamus (a region within and below the thalamus), and there are two types of damage and resulting behavior; if the center (*ventromedial nuclei*) is damaged the animal overeats (*hyperphagia*), and if the side (lateral) is damaged the animal does not eat (*aphagia*) or drink (*adipsia*). There are two phases in hyperphagia. First, an initial *dynamic* phase occurs which follows immediately after the damaging operation. In this phase, the animal eats two or three times as much as normal and gains weight rapidly. After the animal has become obese a *static* phase develops, in which the weight levels off at a high plateau and the animal's food intake drops back to slightly more than normal.

How do hyperphagic animals respond to changes in taste of their diet? Table 2.5 illustrates the change in a five-day intake of food when it is either sweetened or made bitter. Note that a very small percent of quinine has been added in this experiment. It is seen that the diet of the static obese rats has changed rather dramatically with both adulterations, while little change is observed for normal and dynamic phase rats. When hyperphagic animals are fed intragastrically they are very sluggish in adapting, taking seven days to learn to press the bar frequently enough compared to the normal rat's one day. If, however, a very slight taste of food is injected into their mouth with each bar pressing, they learn rapidly and overeat.

TABLE 2.5[a]

Group		Standard diet (g)	0.125% quinine	50% dextrose
1	Normal	16.1	15.8	
	Obese	19.5	2.1	
	Dynamic	25.4	24.4	
2	Normal	16.9		14.2
	Obese	22.2		29.3
	Dynamic	26.5		26.5

[a] From Teitelbaum (1955).

If the experimental animals have lateral hypothalamic lesions they will not be driven to eat by hunger. Only if highly palatable wet foods are offered can they be induced to eat. They must be given wet or liquid foods because they will not drink and otherwise will die of dehydration. With these palatable foods and some force feeding with tubes, the animals eventually recover. As they recover, so does motivation to eat and drink, and taste becomes less important.

Although the origins for the urges to eat and drink are by no means mapped out in the body, some conclusions may be drawn from the above experiments. Normal consumption of nutrients and water depends on adequate motivation and, clearly, taste and smell are strong motivating stimuli. However, in a normal animal they are dispensable. In a brain-damaged animal the normal drive is diminished and the senses become extremely important.

Application of these findings to human motivation is complicated by both genetic and psychological factors. Tests among ethnic groups of specific compounds reveal differences which may be due to sociological factors in their childhood rather than a true genetic factor. Psychological factors are important and milk is an outstanding example. In the Orient, milk is not plentiful and children are taught that it is loathsome (Snapper, 1967), while in the United States a great number of adults drink milk. This taste for milk in the United States probably results from the official recommendation by nutritional authorities (reported in the press to be politically motivated because of farmers' overproduction) that the average daily intake of calcium should never fall below 800 mg. The only convenient way to get this much calcium is by drinking milk. There is little doubt that this recommendation is erroneous since the average calcium intake in other Western countries is 450 mg daily and their health statistics are impressive. However, due to the lay press, TV, and other media, every American citizen is afraid that unless he drinks several glasses of milk a day his bones will stop growing, his teeth will crumble, or his nails will fall out. Since no member of the animal kingdom drinks milk after weaning, the milk drinking habits of Americans must be considered abnormal. What can be wrong with it? With milk comes butter fat, which is highly cholesteric, and is alleged to contribute to the development of arteriosclerosis. Autopsies were performed on American soldiers with an average age of 22 years who were killed in the Korean war. Obstruction in at least one coronary artery was found in 50%, and in 10% one coronary artery was completely closed. In contrast, autopsies on the Korean soldiers showed no arteriosclerosis (Enos *et al.*, 1955). The cause-effect relationship of fat to coronary artery occlusion is not positively established, however. See p. 36.

GUSTATORY IMPULSES IN CORTICAL NEURONS

We have seen that the bonding of a chemical stimulus to a gustatory cell within a taste bud apparently supplies sufficient energy to start an action potential signal through the myelinated nerve fibers. Many gustatory cells are within a single bud, and electron micrographs show that they are interconnected through synapses within the bud (Figure 2.2). Some combination, or blend, of signals goes through a single nerve fiber contained in the bundle of fibers, called the chorda tympani, up to the thalamus, and from there to the limbic lobe for data processing. The experiments described show that all single gustatory cells respond electrically, to a greater or lesser degree, to all four primary tastes. Measurements in single fibers also show that no taste differentiation has taken place. In the search to determine how an organism distinguishes between tastes, the next place to look is within a single nerve cell in the brain which has received the signal. This is done by inserting a microelectrode into the brain and flowing solutions over the tongue. When the electrode is within a single neuron, bursts of electrical impulses are measured. Although this type of experiment has been performed by many, a typical result of Makous *et al.* (1963) will be adequate for our purposes. Figure 2.16 shows the result for a single neuron in the medulla (brain base near the spinal column) of a rat. The ordinate is the number of electrical impulses per second received by the neuron in the first 5 sec of the tongue stimulation, while the lower abscissa is the log of the concentration. It is seen that this single neuron has received signals from NaCl, KCl, and sucrose, although in different magnitudes. Between each concentration a water rinse was applied, which gives the

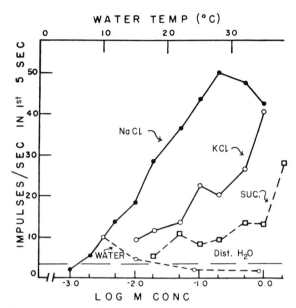

FIG. 2.16 The action potential frequency of a single taste neuron in a rat medulla to chemical and thermal stimulation of the tongue. The water temperature axis refers only to the curve labeled water. [From Makous *et al.* (1963).]

distilled H_2O base line. The dashed water line measures the temperature effect of water, with temperature for the water data given on the upper abscissa of the figure.

How does the brain determine what has been applied to the tongue? This figure indicates that any given frequency of impulses could be caused by any of the three compounds tested at their respective concentrations. Many neurons were measured in this way, and each gave a different set of curves with no discernible information pattern. Clearly, there is not a one-to-one signal of the four tastes from the tongue to the brain, and yet in some way the brain sorts out this apparent hodgepodge of information. It is believed that the brain does this by a pattern recognition process; a simplified version of this is described in the next section.

GUSTATORY NEURAL PATTERNS

A pattern recognition method is independent of the stimulus, that is, it could equally well apply to the information processing of any of the sensory stimuli. It is probably most difficult to apply to taste because, as

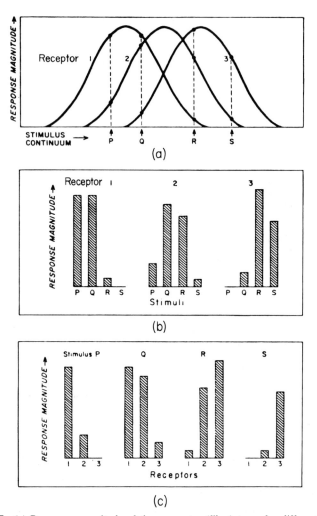

FIG. 2.17   (a) Response magnitude of three receptor (fiber) types for different stimuli. P, Q, R, and S represent four stimuli and vertical dashed lines give their intersections. (b) Magnitude of the response of each of the three receptors to the different stimuli. (c) Pattern effect of the different stimuli on the three receptors. [From Erickson (1963).]

the data in this chapter indicate, the basic taste stimuli are only broad generalizations and details such as the anion effect with the sodium ion are simply unknown. Erickson (1963), in his discussion of the process, prefers to relate it to three color visual perception. We do not wish to describe this process in detail so we will simply assume that there are three classes of tastes, each with a fiber to carry a signal when this taste class has been stimulated. Within a taste class there is probably a broad spectrum of responses, as shown in Fig. 2.9 for some sodium salts. There is undoubtedly some overlap between the types of taste since pH and degree of dissociation have an effect. Erickson plots these three taste spectra in Fig. 2.17. This plot indicates overlap, and it should be recalled that the signal from each curve, or spectra 1, 2, and 3, has its own fiber to signal the brain. If a stimulus is applied to the tongue which has an effect P, the response magnitude of each fiber is that of the intersection of the vertical dotted line above P. The magnitude for the P stimulus is shown in Fig. 2.17b for each of the three fibers (or receptors); similarly with stimuli Q, R, and S. Thus, Fig. 2.17b represents bar graphs from each of three fiber types when standard stimuli P, Q, R, and S are used. This is the type of data obtained on both single neuron measurements, as in Fig. 2.16, and single chorda tympani fibers. Such data plotted as bar graphs are shown in Figs. 2.18a and b. These figures show that there are many fiber types involved in taste, each type having its own characteristic bar graph. These fibers probably form patterns of activity by interconnection (across fiber) between groups of fibers. For example, refer to Fig. 2.17c. Here the grouping is arranged to show how differences in a pattern can be formed by each type of stimulus. The bar graph response of the group 1, 2, and 3 receptors (fibers) is shown for each of the four different stimuli P, Q, R, and S. Thus each stimulus has a recognizable pattern.

The reasonableness of this pattern recognition model was demonstrated in the following way. First, Erickson obtained an across-fiber pattern for three taste solutions for 13 single fibers in the chorda tympani of a rat. This is shown in Fig. 2.19, in which the NH_4Cl and KCl points are connected by lines and the NaCl unconnected for clarity. This figure contains the same data as would 13 bar graphs similar to Figs. 2.18a and b, but is for only three solutions. It is seen in this figure that the NH_4Cl and KCl solutions have a similar response (the product moment of the two curves is quite close), while the NaCl is quite different. In a test on rats, when they were taught to avoid either KCl or NH_4Cl by electrical shocks, they avoided the other also and drank the NaCl. Thus, the experiment demonstrated what the pattern recognition theory predicted from the data of Fig. 2.19; although the various fibers involved in taste sensitivity have considerable diversity and prevent easy fiber classification, across-fiber patterns appear to be the operative identification mechanism and may permit further understanding of the process without detailed knowledge of each fiber's characteristics.

The well-known law of Fechner, namely, that the sensation intensity S is proportional to the logarithm of the stimulus intensity I, is usually expressed as $S = k \log I$. The range of validity of this law will be discussed in some detail in Chapter 7. It is interesting, as we proceed through the

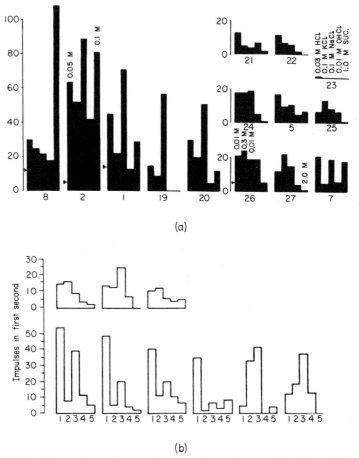

(a)

(b)

FIG. 2.18   (a) Receptor profiles analagous to those shown in Fig. 2.17b obtained from single neurons in a rat. Bar heights indicate number of impulses recorded in the first second of evoked activity. Small triangles indicate spontaneous level of activity. (b) Same as (a) except data obtained from single fibers in the chorda tympani. Stimuli: column 1, 0.1M NaCl; column 2, 0.3M KCl; column 3, 0.03M HCl; column 4, 0.01M QHCl; column 5, 1.0M sucrose. [From Erickson (1963).]

chapters on the senses, to note how this law was empirically established for each. Pfaffman (1963) and his associates compared the signal intensity in the chorda tympani of two groups of rats whose tongues were exposed to different molarities of NaCl solutions. Their experiments showed that saline concentrations above 0.0018 M produce increased discharge. A plot of the results of 3-min stimuli followed by 3-min intervals is shown in Fig. 2.20. For concentrations above threshold and below saturation the data do

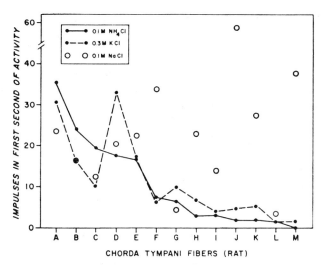

FIG. 2.19 Impulses in 13 (A–M) single fibers of the chorda tympani of a rat for solutions of 0.1M NH$_4$Cl, 0.3M KCl, and 0.1M NaCl on the tongue. [From Erickson (1963).]

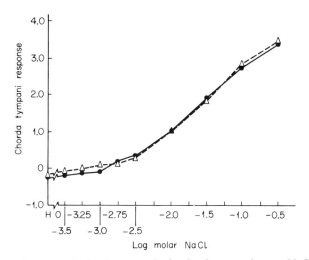

FIG. 2.20 Relative magnitude of response in the chorda tympani versus NaCl concentration on the tongues of two groups of rats. [From Pfaffman (1963).]

obey the relation $S = k \log I$, where S is the magnitude of the chorda tympani response and I is the NaCl concentration; k, of course, is an arbitrary constant. The other senses follow this law, and speculation as to the biophysical reasons will be discussed in Chapter 7.

BIOSTATISTICS AND THE INDICTMENT OF CHOLESTEROL

It is well known that when taking data to prove a point, there is often an unconscious bias in the data selection. For this reason, the Food and Drug Administration requires "double blind" tests of all new drugs. In this type of test some of the patients in the sample are given placebos, while the remainder are given the drug. But neither the patient nor the administering physician knows who is given which, hence the name double blind.

It is an interesting exercise to delve into older statistical data and examine correlations which were never intended to be even part of the original study, since these would not only be double blind but probably completely blind and therefore absolutely free from bias. Some interesting examples of these correlations were reported by Morowitz (1975) in a reexamination of the Hammond Report and will now be discussed.

A three-year study, ending in 1963, was conducted to find correlation between smoking and death rate, and it is called the Hammond Report on "Smoking in Relation to Mortality and Morbidity." Almost 70,000 volunteer workers gathered data on 1 million subjects and followed the records of 400,000 men for three years. As is well known, death rate per age group, or Age-Standardized Death Rate (ASDR), was found to be far higher in male cigarette smokers than in male nonsmokers. In order to eliminate other possibilities, a large number of facts about the life style of the subjects was also taken. In the following tables all data are for nonsmokers.

Table 2.6 is the age–standardized death rate as a function of sleeping patterns. The results are quite clear. Seven hours sleep per night is the healthiest and sleeping less or more is unhealthy. No explanation exists for these data because the study has not been adequately pursued. Morowitz points out, however, that the alarm clock is clearly an underutilized therapeutic tool.

If data on mortality rate versus height is examined, Table 2.7 can be constructed. Individuals with the longest lifetimes are those of height 6 ft–6 ft 1 in. Again, no reason exists nor is there a readily available remedy.

TABLE 2.6

Average hours sleep per night	ASDR
less than 5	2029
5	1121
6	805
7	626
8	813
9	967
10 or more	1898

TABLE 2.7

Height in inches	ASDR
less than 66	1065
66–67	815
68–69	806
70–71	784
72–73	687
74 and over	735

When education is examined the following results in Table 2.8 are obtained. Clearly, the more education the longer lived is the individual. One can begin to suggest reasons such as better knowledge of health and nutrition, or possibly a higher income bracket permits better practice of both.

TABLE 2.8

Education	ASDR
Grammar school or less	945
Some high school	864
High school graduate	766
Some college	755
College graduate	676

The most startling finding, however, is the correlation with consumption of fried foods. This is shown in Table 2.9. The correlation is clear. Except for a possible optimum of three to four times a week, the death rate decreases with increasing consumption of fried foods. There was no apparent bias in obtaining the correlation since these data were from a study on smoking habits. These results are startling because of the Framingham study on cholesterol and death rate which has resulted in the Surgeon General's warning against consumption of fried foods.

TABLE 2.9

Subgroup	ASDR
No fried foods eaten	1208
Fried food 1–2 times a week	1004
Fried food 3–4 times a week	642
Fried food 5–9 times a week	781
Fried food 10–14 times a week	722
Fried food 15 + times a week	702

A ten-year study was done on 5000 residents of Framingham, Massachu-setts, again following all habits and death rates. The study indicted choles-terol, a basic component of fatty and fried foods, as the common factor—the greater the cholesterol consumption the greater the death rate. This was believed to be the cause of fatty deposits in the arteries, particularly the coronary arteries, which leads to an insufficient blood supply to the heart muscle. The result of the study was not unexpected because of the prior suspicion that, since there were often fatty deposits in the coronary arteries of patients with heart attacks, the consumption of fat must be involved. (See p. 30.)

At the time, the correlation was obvious and clear and physicians, including the Surgeon General, advocated less fat consumption for a longer life. However, there have been some dissonant voices saying that the correlation is too simplistic. Recent research has indicated some new factors which must be studied and considered. For example, the body manufactures cholesterol and, if one does not ingest a sufficient quantity for the development and replacement of the cell membranes which require it, the body will make up the difference. It has been found that cholesterol transport is governed by two lipoproteins, one of heavy and one of light molecular weight. The light lipoprotein delivers the cholesterol to the cells and the heavy lipoprotein returns the excess to storage. Some individuals have a faulty ratio, that is, they lack sufficient heavy lipoproteins, and the assumption is that in these individuals a buildup of fatty deposits can occur. A recent study of people who jog or run indicates that such vigorous exercise can favorably alter the ratio by increasing the amount of heavy lipoproteins in the blood stream.

The cause of the deposit of the excess cholesterol in the arteries seems to be related to pathological lesions in the arterial walls. Recent studies on chickens have shown that if they are infected by a herpesvirus of fowl they will develop atherosclerosis whether they are fed a cholesterol-free or a cholesterol-supplemented diet, although those fed an excess of cholesterol develop larger and more numerous deposits. Pathogen-free chickens do not develop atherosclerosis on either diet [*Chem. and Eng. News* **58**, 7 (1980)]. Humans are susceptible to a variety of herpesviruses, and once a person acquires the infection it is usually permanent. Thus, if an inoculation against herpesvirus is developed, atherosclerosis may be eliminated from the population independent of diet practice. The blood platelets also have been shown to play a significant role in the development of atherosclerosis, although all of the pieces of the puzzle have not yet been put together. (Mustard *et al.*, 1977).

Clearly, the Framingham conclusions involve a philosophy similar to the discussion of salt and hypertension earlier in this chapter. Because some of the population is genetically predisposed to hypertension, a reduction of salt intake for all becomes a prophylaxis. Because some of the population may be both infected with herpesvirus and have an inadequate supply of heavy lipoproteins, a reduction in fat intake for all is a prophylaxis, whether they need it or not. Table 2.9 suggests that a large fraction of the population is either not infected or has an adequate lipoprotein balance.

There is a moral to be found in this story of the indictment of cholesterol, but it is premature to state it with any certitude.

Note added in proof: Since this was written two reports have also questioned the cause–effect relation of cholesterol consumption and heart disease. [*Harvard Medical School Health Letter* V (10) 6 (1980); "Toward Helpful Diets," Food and Nutrition Board of the National Academy of Science, 1980].

REFERENCES

Beidler, L. M. (1953). Properties of chemoreceptors of tongue of rat, *J. Neurophysiol.* **16**, 596.

Beidler, L. M. (1954). A theory of taste stimulation, *J. Gen. Physiol.* **38**, 133.

Beidler, L. M. (1962). Taste receptor stimulation, *Prog. Biophys. Biophys. Chem.* **12**, 109.

Beidler, L. M. (1970). Taste bud cells act as receptors and not merely as chemical filters, *in* "Oral Sensation and Perception" (J. F. Bosma, ed.). Thomas, Springfield, Illinois.

Borg, G., Diamant, H., Oakley, B., Ström, L., and Zotterman, Y. (1967). A comparative study of neural and psychophysical responses to gustatory stimuli, *in* "Olfaction and Taste" (T. Hayashi, ed.), Vol. II. Pergamon, Oxford.

Bull, T. R. (1965). Taste and the chorda tympani, *J. Laryng. Otol.* **79**, 479.

Dahl, L. K. (1958). Salt intake and salt need, *New England J. Med.*, **258**, 1152, 1205.

Dahl, L. K. (1960). Salt, fat and hypertension: the Japanese experience, *Nutrition Rev.* **18**, 97.

Dahl, L. K. (1967). Effects of chronic excess salt ingestion-experimental hypertension in the rat: correlation with human hypertension, *in* "Epidemiology of Hypertension" (*Proc. Int. Symp. Epidemiol. Hypertension, Chicago, Illinois, 1964*). Grune and Stratton, New York.

Dahl, L. K., Heine, M., Leitl, G., and Tassinari, L. (1970). Hypertension and death from consumption of processed baby food by rats, *Proc. Soc. Exp. Biol. Med.* **133**, 1405.

Diamant, H., Oakley, B., Ström, L., Wells, C., and Zotterman, Y. (1965). A comparison of neural and psychophysical responses to taste stimuli in man, *Acta Physiol. Scand.* **64**, 67.

von Ebner, V. (1899-1902). Von den Verdanungsorgan, *in* "Koelliker's Handbook der Gewebelehre des Menschen," Vol. 3, p. 1. Englemann, Leipzig.

Enos, W. F., Beyer, J. C., and Holmes, R. H. (1955). Pathogenesis of coronary disease in American Soldiers killed in Korea, *J. Am. Med. Assoc.* **158**, 912.

Epstein, A. N., and Teitelbaum, P. (1962). Regulation of food intake in the absence of taste, smell and other oro-pharyngeal sensations, *J. Comp. Physiol. Psychol.* **55**, 753.

Erickson, R. P. (1963). Sensory neural patterns and gustation in olfaction and taste, *in* "Olfaction and Taste" (Y. Zotterman, ed.). Pergamon, Oxford.

Erickson, R. P. (1967). Neural coding and taste quality, *in* "The Chemical Senses and Nutrition" (M. R. Kare and O. Maller, eds.). Johns Hopkins Press, Baltimore, Maryland.

Fishman, I. Y. (1957). Single fiber gustatory impulses in rat and hamster, *J. Cell Comp. Physiol.* **49**, 319.

Heidenhain, M. (1914). Über die Sinnesfelder und Geschmacksknopfen der Papille Foliata des Kaninchens, *Arch. Mikr. Anat.* **85**, 365.

Henkin, R. I. (1970). The role of unmyelinated nerve fibers in the taste process, *in* "Oral Sensation and Perception" (J. F. Bosma, ed.). Thomas, Springfield, Illinois.

Kare, M. R., and Maller, O. (eds.) (1967). "The Chemical Senses and Nutrition." Johns Hopkins Press, Baltimore, Maryland.

Kimura, K., and Beidler, L. M. (1961). Microelectrode study of taste receptors of rat and hamster, *J. Cell Comp. Physiol.* **58**, 131.

Kionka, H., and Strätz, F. (1922). Does the taste of salt depend upon the tastes of the individual ions or upon the entire molecule? *Arch. Exp. Path. Pharmak.* **95**, 241.

Kitchell, R. L. (1963). Comparative anatomical and physiological studies of gustatory mechanisms, *in* "Olfaction and Taste" (Y. Zotterman, ed.). Pergamon, Oxford.

Makous, W., Nord, S., Oakley, B., and Pfaffman, C. (1963). The gustatory relay in the medulla, *in* "Olfaction and Taste" (Y. Zotterman, ed.). Pergamon, Oxford.

McBurney, D. H., and Pfaffmann, C. (1963). Gustatory adaptation to saliva and sodium chloride, *J. Exp. Psychol.* **65**, 523.

Moncrieff, R. W. (1967). "The Chemical Senses," 3rd ed. CRC Press, Cleveland, Ohio.

Morowitz, H. J. (1975). Hiding in the Hammond Report, Hospital Practices, August.

Mustard, J. F., Moore, S., Packham, M. A., and Kinlough-Rathbone, R. L. (1977). Platelets, thrombosis and atherosclerosis, *Prog. Biochem. Pharmacol.* **13**, 312.

Pfaffman, C. (1963). Taste stimulation and preference behavior, *in* "Olfaction and Taste" (Y. Zotterman, ed.). Pergamon, Oxford.

Pyke, M. (1944). Nutrition and a matter of taste, *Nature* (*London*) **154**, 229.

Sato, T. (1980). Recent advances in the physiology of taste cells, *Neurobiology*, **14**, 25.

von Skramlik, E. (1926). *Handbuch Physiol. Nied. Sinne* **1**, 7.

Snapper, I. (1967). The etiology of different forms of taste behavior, *in* "The Chemical Senses and Nutrition" (M. R. Kare and O. Maller, eds.). Johns Hopkins Press, Baltimore, Maryland.

Teitelbaum, P. (1955). Sensory control of hypothalamic hyperphagia, *J. Comp. Physiol. Psychol.* **48**, 156.

Teitelbaum, P., and Epstein, A. N. (1962). The lateral hypothalmic syndrome: recovery of feeding and drinking after lateral hypothalmic lesions, *Psychol. Rev.* **69**, 74.

Teitelbaum, P., and Epstein, A. N. (1963). The role of taste and smell in the regulation of food food and water intake, *in* "Olfaction and Taste" (Y. Zotterman, ed.). Pergamon, Oxford.

Zotterman, Y. (1963). "Olfaction and Taste." Pergamon, Oxford.

CHAPTER 3

Olfaction

INTRODUCTION

The sense of smell, called olfaction, was discussed in the Introduction in Chapter 2. In brief, the olfactory system is the most primitive one which is not differentiated from taste in sea animals, where it is a chemical detector responsive to waterborne molecules and ions. With the evolutionary development of land animals, two separate chemoreceptor systems had to develop: olfaction, which was responsive to airborne molecules, and taste, which was responsive to molecules in solution or, if they were not in solution, they were first dissolved by saliva. Because on land the air was the carrier of most information such as food, danger, or sex, the olfactory system became more refined in its discrimination although, like taste, there is a reasonable number of types of responses, and smells can be classified as combinations of this base number.

Man has lost the olfactory sensitivity that other members of the animal kingdom have; consider the tracking ability of a bloodhound or the few number of molecules required as a sex attractant for insects. The bilateral

41

arrangement of two nostrils probably serves as a locating device in lower animals, just as two eyes and two ears do.

Again, we face the difficulty of performing experiments on animals and inferring the mechanism in man. Instead of washing the tongue with a known molar solution, air with a known concentration of odorous molecules is blown in the nostrils and either electrical signals are recorded in animal neural systems or subjective responses are obtained from humans. Attempts are then made to understand and correlate these experiments. We will see that the chemical equilibrium model of stimulus–receptor interaction developed for taste is applicable to olfaction, and that there has been considerable progress in understanding why certain molecules have a similar smell. The precise mechanism of how the molecule, when it interacts with a receptor, can signal the brain is still unknown.

Some general properties of odorant molecules can be stated from simple considerations of the process. (1) The substance must be volatile so that the molecules may be airborne. (2) The molecules must be somewhat soluble in a mucus–water mixture so that they can be absorbed in the layer which covers the olfactory epithelium. (3) They must also be somewhat soluble in lipids (fatty substances) so that they can penetrate the lipid layer of the membrane that surrounds the nerve ending. (4) They must have an appropriate configuration (electronic or steric) to attach to the receptor molecule and subsequently alter its configuration in such a way that the resting electric potential is changed sufficiently to induce a nerve impulse.

ANATOMY

Only part of the nose is visible. The interior extends inward for $2\frac{1}{2}$ –3 in. where it joins the top of the throat, or pharynx. The nose has many convoluted passages which the air must follow, Fig. 3.1. These cause turbulence in the flow which allows dust to stick to the mucus in the passages, and, furthermore, the long flow path facilitates warming of the air and, possibly, cooling of the brain. Figure 3.2 shows a cross section of the nasal cavities, frontal view, in which the scroll-like structures of the air passages are evident.

In an early (1882) experiment to locate the pathway of the air, E. Paulsen sawed the head of a cadaver in two and placed red litmus paper in various parts of the nasal cavities. He then put the two halves together and blew ammonia through the nose. Upon taking the two halves apart, he could follow the course of the air by the path of the blue litmus paper. Somewhat more sophisticated experiments followed this pioneering work, but all have shown that the olfactory cleft is not in the air pathway. A smell is sensed when breathing in or out through the nose, but not when breathing is stopped. This suggests that there is a turbulence-assisted

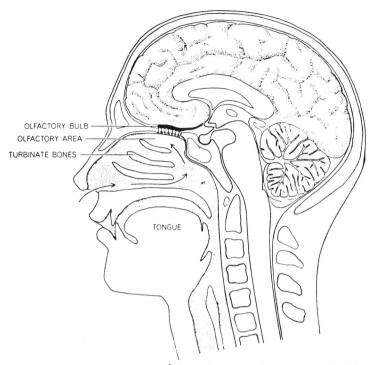

FIG. 3.1 Anatomy of the nose and olfactory region. [From J. E. Amoore, J. W. Johnston, Jr., and M. Rubin, The stereochemical theory of odor, *Sci. Am.* **210** (Feb.), 42 (1964). Copyright © 1964 by Scientific American, Inc. All rights reserved.]

FIG. 3.2 Section across the nasal cavities showing the scroll-like structure of the turbinates. 1, molar teeth; 2, antrum; 3, interior turbinate; 4, middle turbinate; 5, superior turbinate; 6, olfactory cleft; 7, inferior meatus; 8, middle meatus; 9, superior meatus; 10, nasal septum. [From Moncrieff (1967).]

diffusion of a small sample when air is moving past the opening to the olfactory cleft.

While the main interior of the nasal cavities is reddish, the color of the tissue of the olfactory region is yellow. In an adult, the total area in each of

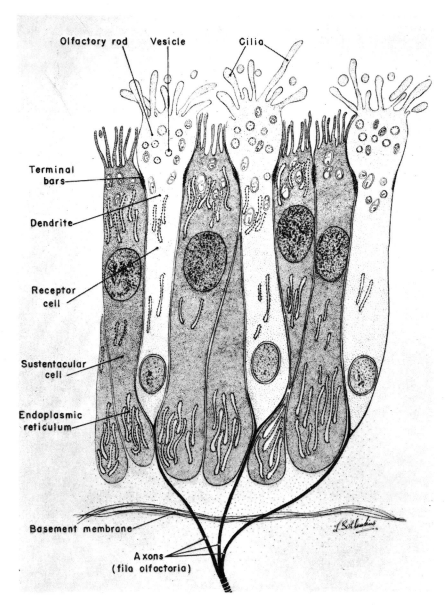

FIG. 3.3 Schematic representation of olfactory mucosa showing olfactory receptor rods and sustentacular cells based on electron microscope studies. [From DeLorenzo (1963).]

the two nasal chambers is about 1 in². This tissue, called the *olfactory mucosa*, has millions of receptor cells supported by *sustentacular cells*. A schematic drawing from an electron micrograph is shown in Fig. 3.3. The upward extension of the receptor cell is called the *olfactory rod*, and it has 6 to 12 small filamentary extensions called *cilia* on which the olfactory receptor sites are located. Spaced between the olfactory cells are many small *glands of Bowman* (Fig. 3.4), which secrete mucus onto the surface of the olfactory membrane.

The axons at the base of the receptor cell are myelinated fibers, but the olfactory rod and cilia are bare or "unsheathed." The nerve fibers, about

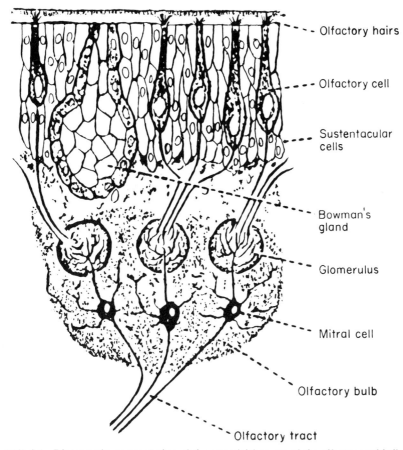

Olfactory hairs

Olfactory cell

Sustentacular cells

Bowman's gland

Glomerulus

Mitral cell

Olfactory bulb

Olfactory tract

FIG. 3.4 Diagramatic representation of the essential features of the olfactory epithelium based on electron micrographs. The height has been shortened. [From Guyton (1971).]

0.2-μm diameter, are grouped into larger bundles which go to the olfactory region of the brain. Note that there are no synaptic junctions between the olfactory cells and the receiving part of the brain, called the *olfactory bulb*.

The nerve axons penetrate into the olfactory bulb where they make many branches and entanglements with dendrites of triangularly shaped *mitral cells*, Fig. 3.4. These regions of entanglements are in spheroidal clusters called *glomeruli*, where synapse takes place. Approximately 25,000 axons enter the glomerulus region and synapse with only 24 mitral cells. The signals from these mitral cells are then sent to the brain. There are several pathways from the mitral cells, the two principal ones being to the *medial olfactory area*, which is in the midportion of the brain above and in front of the hypothalamus, and to the *lateral olfactory area*, which is slightly above the *hippocampus*. Secondary olfactory paths go from both of these regions into the *hypothalamus, thalamus, hippocampus*, and *brain stem nuclei*, and thus control the automatic responses of the body to olfactory stimuli.

The lines of communication have been estimated for a rabbit (Moncrieff, 1967). On each side of the head there are:

(1)   Fifty million receptors, each connected by a separate fiber (primary neurons) to the olfactory bulb.

(2)   In the bulb there are 1900 glomeruli, from each of which there are 24 fibers or secondary neurons, making $1900 \times 24 = 45,600$ which pass to the cerebrum.

If there are 24 different kinds of fibers and each transmits a different kind of primary odor, then any given odor will stimulate a fraction of these 24 kinds into either an "on" or "off" signal (stimulated or not stimulated). The number of variations of the signal to the brain is 2^{24}, or over 16 million kinds of odor are, in principle, detectable.

ELECTRICAL IMPULSES IN THE OLFACTORY SYSTEM

Even though there are no synapses to premix the signals from olfactory receptors, as in the taste receptors, the electrical signals are proving to be much more difficult to unravel. Although there is no shortage of electrical impulses, the major difficulty lies in classifying them in order of importance. In early investigations, Adrian (1950) inserted electrodes into the olfactory bulbs of experimental animals and subsequently subjected them to various odors. He showed that not only were electrical changes observable but that there is a rhythmic electrical pulsation superimposed, which is characteristic of brain activity.

In a specific experiment on the olfactory mucosa, Ottoson (1956) placed an electrode in contact with the olfactory bulb while passing odors through the nostrils of the test animal. He observed a negative change in the electrical potential of a few millivolts, the magnitude apparently depending on the stimulus intensity up to a saturation level. This potential change had a rise time of about 1/2 sec followed by an exponential decay of several seconds. An example is shown in Fig. 3.5, in which the negative voltage is plotted in the positive direction. He called such curves electro-olfactograms (EOG), and more detailed measurements are usually seen superimposed on the EOG. It is thought that the EOG is a lowering of the resting potential which permits the sending of action potential signals. It is speculated that upon stimulation the glands of Bowman secrete mucus with sufficient ions to short-circuit the cells after a few seconds.

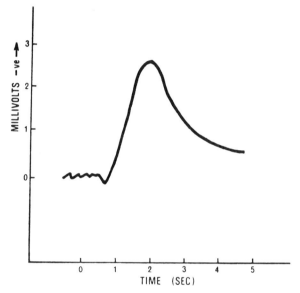

FIG. 3.5 Electro-olfactogram (EOG) for a frog's olfactory mucosa stimulated with butanol vapor. Negative voltage is plotted in positive direction. [From Ottoson (1956).]

If short time pulses are measured along with the EOG within one or two olfactory receptors, a series of spikes, or action potentials, can be seen. The number of spikes depends on the strength of the stimulus, as does the height of the EOG. An example of this is illustrated in Fig. 3.6. This is a recording from a microelectrode inserted into a single olfactory receptor cell of a frog with increasing strengths of airborne n-butanol, the greatest at the bottom. When the position of the probe was changed, the electrode tip would contact a different cell, sometimes two or more cells. Gesteland et al. (1963) observed that the responses of the different cells varied with the type of stimulus. One example is shown in Fig. 3.7. This cell responds to musk xylene, less so to nitrobenzene, slightly to benzonitrite, and not at all to pyridine. These investigators tested approximately 30 cells with 26 different chemicals and recorded the response. The complexity is formidable, but apparently not insoluble. The situation seems to be like that of taste, in which there is not a unique response of a receptor but, rather, many possible variations within a type.

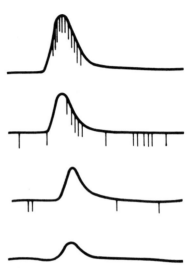

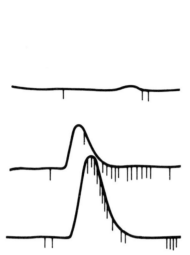

FIG. 3.6 Comparative responses of a single frog olfactory receptor cell to increasing stimulus strengths of n-butanol (bottom curve maximum stimulus). [Redrawn from Gesteland et al. (1963).]

FIG. 3.7 A receptor cell response to (top to bottom) musk xylene, nitrobenzene, benzonitrite, and pyridine. [Redrawn from Gesteland et al. (1963).]

CHEMICAL EQUILIBRIUM MODEL OF OLFACTORY INTENSITY

The olfactory receptor cells are connected through very fine fibers to the olfactory bulb of the brain. There are no synapses in between, a situation which might be expected to simplify the understanding of the electrical impulses. This is because there is generally not a one-to-one transmission of fiber impulses through a synapse to another fiber. For example, as previously discussed, there are, in a rabbit, millions of primary fibers for every postsynaptic fiber connecting with other parts of the central nervous system. Thus, an experiment, in principle, can detect the impulse from a single receptor cell in a single nerve fiber. Unfortunately, the fibers are so small that meaningful experiments on a single fiber are difficult. The experiments and interpretation of Tucker (1963) on "twigs" of fibers leading from the olfactory mucosa of tortoises will now be described.

The receptors were exposed to controlled flow rates of known concentrations in air of various chemical compounds. Upon exposure, oscilloscope

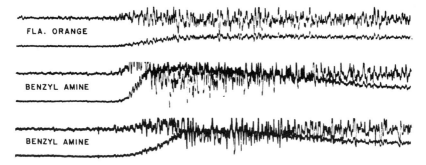

FIG. 3.8   Oscilloscope traces of electrical activity in a nerve twig of a tortoise when exposed to a brief pulse of odorous air. The ac trace above and dc trace below in each pair. [From Tucker (1963).]

traces of the resulting electrical impulses appear, as shown in Fig. 3.8. In each pair of traces, the upper one is an ac amplifier trace which shows the action potential, or "spike," activity, while the lower dc trace shows the EOG potential. This spike activity was then put through an integrating circuit which recorded the total electrical activity.

Figure 3.9 illustrates the effect of adaptation of the olfactory receptors and their recovery. In the upper trace, amyl acetate flows through the nose for 40 sec with clean air in between. From left to right the concentrations,

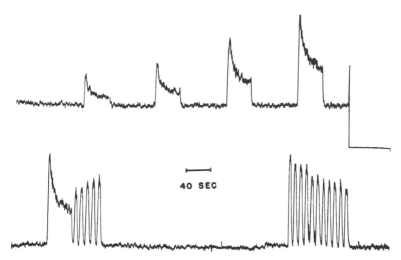

FIG. 3.9   Electrical activity of olfactory response to amyl acetate. Upper trace: 40-sec exposure followed by clean air rinse for increasing concentrations of 10^{-3}, $10^{-8/3}$, $10^{-7/3}$, 10^{-2} fraction of saturation. Lower trace: left side, 40 sec on followed by 5 sec on and off; right side, 5 sec on and off. [From Tucker (1963).]

expressed in fractional saturation of the air, are 10^{-3}, $10^{-8/3}$, $10^{-7/3}$, and 10^{-2}. These curves show a rise time of a few seconds, presumably while the chemical diffuses through the mucous layer to the receptor, followed by a decay of activity with about a 10-sec time to half-value as the receptors adapt to the presence of the odor. In the left-hand part of the lower trace, the flow of amyl acetate of 10^{-2} of saturation is on for 40 sec, followed by

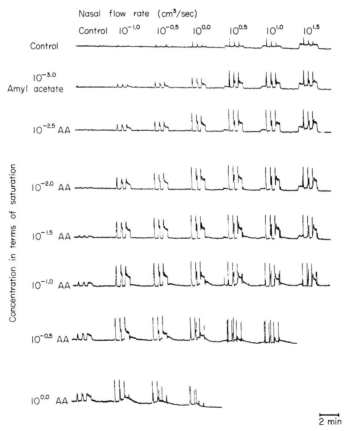

FIG. 3.10   Integrated electrical activity from olfactor receptors of a tortoise as a function of concentration of amyl acetate and flow rate. Each group of peaks from left to right indicates the activity due to 5-, 10-, and 30-sec exposure with 30-sec clean air rinse between the 5- and 10- and between the 10- and 30-sec exposures. [From Tucker (1963).]

alternations of 5 sec on and 5 sec off. Notice here that the sensitivity is being recovered. In the right-hand part of the lower trace, the 5 sec on and off method is used without the prior 40-sec exposure. Note that adaptation is occurring as the electrical activity decreases with each flow pulse.

Figure 3.10 shows a group of data similar to that of Fig. 3.9. Amyl acetate of the indicated concentrations flows at different rates past the receptors. The data are shown as groups of three peaks. In each group, the time at which the flow is on, and therefore the duration of the peak, is 5, 10, and 30 sec, from left to right. In between each of these odor periods is a 30-sec clean air rinse. It is evident from this figure that the electrical activity increases with both concentration and flow rate, up to a saturation value.

The magnitudes of the olfactory responses to amyl acetate of Fig. 3.10 are plotted in Fig. 3.11, on the left as a function of concentration and on the right as a function of flow rate. In the right-hand figure the grouping together of the four highest concentrations at the higher flow rates indicates that increasing the flow rate increases the odorant concentration at the receptors until the concentration can go no higher. In other words, inhaling a strong odor more deeply will not make the smell more intense past a certain level. In addition to this effect, the right-hand figure also indicates that for any concentration there is a receptor saturation at a given flow rate, in the present case all of the curves level off at about 4 cm^3/sec, and that the slopes prior to saturation are all about the same. Therefore, the curves can all be displayed vertically and will superimpose within experimental error. This means that if we look at the half-value of activity, it occurs for all the curves at the same flow rate. It may therefore be concluded that the concentration at the receptors appears to be a constant fraction of the inlet concentration for a given flow rate. If, in addition to the half-value of electrical response in the right-hand figure, we take other vertical cuts at different flow rates through the curves, we would know the values of the concentrations at the receptors for each of the flow rates. The left-hand figure is just such a series of curves in which the actual data points rather than the theoretical curve intersection points are used. But for all the data inaccuracies, the curves in the left-hand figure still represent *known* values for the concentrations at the receptors. Since the curves on the left also represent theoretical values of concentration at the receptors, we can translate the lower flow rate curves to the left and they should superimpose on the upper curves. Such an operation has the additional advantage of giving response as a function of concentration at the receptors below the range available from the experiment. When such a translation is done, the data point positions of Fig. 3.12 are obtained.

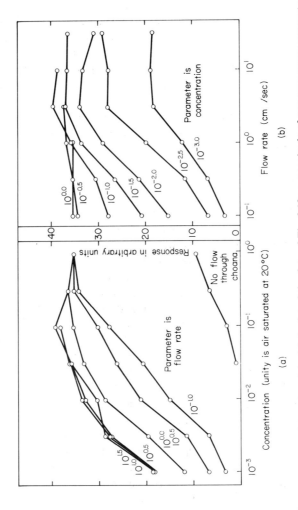

FIG. 3.11 Olfactory response to amyl acetate. (a) A plot of the data of Fig. 3.10 as magnitude of response versus concentration for different flow rates. (b) A plot of the data of Fig. 3.10 as magnitude of response versus flow rate for different concentrations. [From Tucker (1963).]

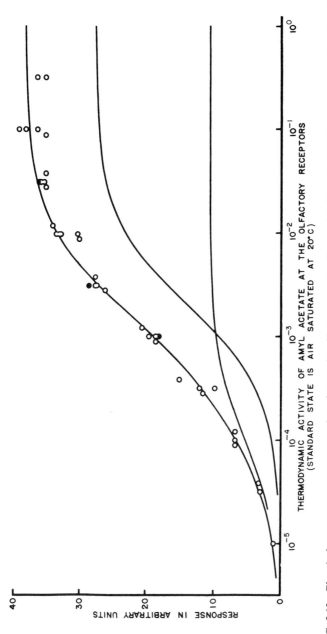

FIG. 3.12 Electrical response versus concentration of a tortoise olfactory receptor. Data points from Fig. 3.10, lower curves are from Eq. (3.2) and the upper curve is the sum of the two lower curves. [From Tucker (1963).]

THERMODYNAMIC ACTIVITY OF AMYL ACETATE AT THE OLFACTORY RECEPTORS
(STANDARD STATE IS AIR SATURATED AT 20°C)

RESPONSE IN ARBITRARY UNITS

To interpret these data, Tucker followed the chemical equilibrium model developed by Beidler for taste receptors, which is described in Chapter 2.

$$\text{stimulus} \quad + \quad \text{receptor sites} \quad \overset{K_2}{\underset{K_1}{\leftrightarrows}} \quad \text{stimulus–receptor}$$

$$C \quad + \quad N - Z \quad\quad\quad\quad Z$$

where C is the concentration of stimulating molecules, N the concentration of receptor sites, and Z the concentration of stimulus–receptor pairs. Following Eq. (2.1), we write Eq. (2.3)

$$K = Z/C(N - Z) \tag{2.3}$$

assuming that the activity coefficients are near unity, and we can write the association reaction in terms of concentration instead of activities (see Vol. I, Appendix A, Section A.7). Call r the electrical response and assume that it is proportional to the number of stimulus–receptor pairs, Z. Since $0 \leqslant Z \leqslant N$, r has corresponding values from 0 to r_{max}, the latter value occurring when $Z = N$. If k is the constant of proportionality connecting electrical response with concentration of stimulus receptor pairs, then

$$r = kZ \quad\quad \text{and} \quad\quad r_{max} = kN \tag{3.1}$$

Solve Eq. (2.3) for Z and obtain

$$Z = NKC/(1 + KC) \tag{2.3}$$

then substitute the relations of Eq. (3.1) and the response equation may be written as

$$r = r_{max}KC/(1 + KC) \tag{3.2}$$

Tucker then tried unsuccessfully to match the data points of Fig. 3.12 with this equation.[†] However, he recognized that he was measuring the response in a nerve twig, which included the fibers from several receptors, i.e., multiple binding, rather than a single one. It is not expected that the equilibrium constant between a stimulus molecule and all receptors would be the same, and the measured response would therefore be the sum of the

[†] In his paper Tucker (1963) reports that he used the equation

$$r = \tfrac{1}{2} r_{max}(1 + \tanh \tfrac{1}{2} \ln KC)$$

in the curve fitting process. This is the trancendental form for the semilogarithmic plot of Eq. (3.2). He could just as easily have used Eq. (3.2). They are readily shown to be identical by recognizing that

$$\tanh u = (1 - e^{-2u})/(1 + e^{-2u})$$

where $u = \tfrac{1}{2} \ln KC$. But $e^{-2[(1/2) \ln KC]} = 1/e^{2[(1/2) \ln KC]} = 1/KC$ by the identity $e^{\ln x} = x$. Rearrangement of the terms yields Eq. (3.2).

responses of the types of receptors. This total response is the simple sum of Eq. (3.2) for each and can be written as

$$r_{total} = \frac{r_{max\,1}K_1C}{1 + K_1C} + \frac{r_{max\,2}K_2C}{1 + K_2C} + \cdots \tag{3.3}$$

Tucker found that he could match the data with only two types of receptors. The lower curves of Fig. 3.12 are plots of Eq. (3.2) with two different equilibrium constants and r_{max}. The total response, which by Eq. (3.3) is the sum of these two curves, is the line drawn through the data points. This result is a convincing demonstration of the chemical equilibrium model.

Examination of Eq. (3.2) shows that when $K = 1/C$ the response is one-half the maximum, $r = r_{max}/2$. Referring to the two lower curves of Fig. 3.12, it is seen that $r_{max}/2$ occurs around $C = 10^{-3}$ for one and $C = 10^{-4}$ for the other. The two equilibrium constants for the two types of receptors are therefore $K = 10^3$ and $K = 10^4$. Following the discussion of this chemical equilibrium model in Chapter 2,

$$\Delta G^0 = -RT \ln K \tag{3.4}$$

and for the two receptors ΔG^0 has values at room temperature of about -780 and -1040 cal/mole. Tucker also measured the temperature effects of the response over a 10°C range and found none. The same arguments as in taste apply; ΔG^0 is too small to arise from broken and reformed chemical bonds. The attaching force must therefore arise from the entropy and be physical in nature.

THE WEBER–FECHNER RELATION IN OLFACTION

We have shown in Chapter 2 that a plot of the logarithm of perceived intensity versus logarithm of actual stimulus intensity is linear. This is true between threshold and saturation of the particular sensor for all types of stimuli, and such a relationship is called the Weber–Fechner Law. The underlying mechanism for this will be discussed in Chapter 7.

Cheesman and Mayne (1953) obtained such data from human subjects with four odorants. Such data must be taken with care and appropriate statistical analysis. Nevertheless, they are quite believable and the consistency of agreement with the Weber–Fechner relationship lends them considerable credibility.

Plots of the data for ethyl mercaptan, amyl alcohol, acetone, and isopropanol are shown in Fig. 3.13. The equation of the lines can be written as

$$\ln I = k(\ln C - \ln C_t) \tag{3.5}$$

where I is the perceived intensity, C is the concentration of the odorant, C_t is the threshold concentration for perception, and k is a constant, equal in this case to 0.6.

The extreme sensitivity of some receptors can be seen in this figure, particularly for the mercaptan. Note that the formula for ethyl alcohol is C_2H_5OH and for ethyl mercaptan it is C_2H_5SH, that is, the oxygen is replaced by a sulfur atom. If there are about 20 cm^3/sniff of air and the density of air at body temperature is 0.00116 g/cm^3, the weight of the air is 0.023 g. If air is primarily nitrogen molecules of weight 14 g/(6×10^{23}), where the denominator is Avogadro's number, the weight is about 2×10^{-23} g/molecule. Therefore, the number of molecules of nitrogen in the 20 cm^3/sniff is 0.02 g/$(2 \times 10^{-23}$ g/molecule) or $0.01 \times 10^{23} = 10^{21}$ molecules/sniff. Figure 3.13 shows that roughly one molecule in 10^{20} of mercaptan is detectable as an odor and therefore about 10 molecules in a single sniff are detectable.

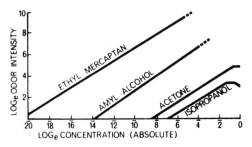

FIG. 3.13   Plot of log of human sensation to four odors as a function of the molecular concentration. [From Cheesman and Mayne (1953).]

It has been said that the number of olfactory theories is almost equal to the number of investigators. Although a modest degree of hyperbole is used, olfactory theories have undoubtedly proliferated because there are so few reliable experiments which can be done to settle the matter. We will discuss two of them, selected because one appears to have a reasonable empirical justification, while the other is an attempt to quantify a mechanism of olfaction. In fact, both of these models, as well as some others not discussed here, may eventually be proven to be different approaches to the correct model.

STERIC MODEL OF OLFACTION

Chemists have long known that certain molecules which look alike have a similar odor. Amoore (1964) and others have pursued this "steric theory" of odor with considerable perserverence and have been able to classify the fundamental odors. After testing many molecules on human subjects, they decided that there were seven basic odors and all others were combinations of these. They are listed in Table 3.1.

TABLE 3.1

Primary odor	Chemical example	Primary odor	Chemical example
Camphoraceous	Camphor	Ethereal	Ethylene dichloride
Musky	Pentadecanolactone	Pungent	Formic or acetic acid
Floral	Phenylethylmethyl ethyl carbinol	Putrid	Butyl mercaptan
Pepperminty	Menthone		

The structures of many molecules were examined, and it was found that molecules with the above classes of odors had certain shapes associated with them. Moncrieff (1967) had proposed in 1949 that there were certain shapes of receptors which correspond to the primary odors and that odorous molecules produce their effects by fitting closely into these sites. This is the same "lock and key" hypothesis that has been useful in explaining the interaction of enzymes with their substrates, antibodies with antigens, and DNA with messenger RNA.

By examining the x-ray structure of odorous molecules, the space-filling forms of the seven primary odors were approximated, see Fig. 3.14. Three of these depend primarily on molecular size. Camphoraceous molecules have a hemispherical basin as a receptor, and globular molecules of about 7-Å diameter fit in and produce this odor. The ethereal site is a narrow slot which accepts only very small or thin molecules. The musky site is a larger, elliptical, flat-bottomed pan which requires a big disk-shaped molecule to fill it adequately. Two of the primary odors result from special molecular shapes rather than size. Floral odor is caused by molecules that have the shape of a disk with a flexible tail attached, somewhat like a kite. The minty site is designed for wedge-shaped molecules. However, there is an additional requirement. The point of the wedge must be capable of forming a hydrogen bond, presumably because there is an electrophilic hydrogen atom in the receptor site at this position. The final two primary odors do not seem to depend on the size or shape of the molecule but on its electronic charge. Pungent molecules, because of a deficiency of electrons, have a net positive charge and therefore a strong affinity for electrons; these molecules are called electrophilic. Putrid odors have an excess of electrons and are strongly attracted to nuclei of adjacent atoms; these are called nucleophilic.

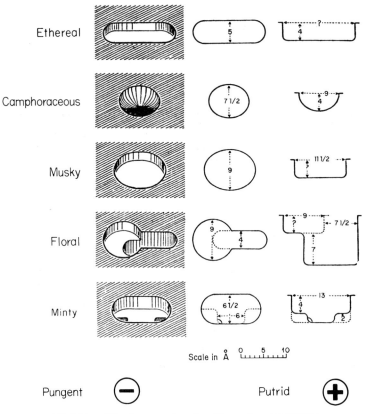

Ethereal

Camphoraceous

Musky

Floral

Minty

Scale in Å

Pungent ⊖   Putrid ⊕

FIG. 3.14a Olfaction receptor sites. [From Amoore (1964).]

This stereochemical model was tested in a number of ways. Molecules of a certain shape were synthesized and their odor predicted. These predictions were confirmed by a panel of trained testers. A number of compounds have the odor of cedarwood, a complex odor found in nature. Amoore found that molecules of compounds with this odor would fit the sites for camphoraceous, musky, floral, and pepperminty odors. Following the reasoning that complex odors were formed from combinations of the primary odors, Johnston (1963) was able to prepare a blend of these primaries which the trained observers could not distinguish from cedarwood. Many similar tests have been performed with the conclusion that this model has merit, nevertheless it remains controversial.

These measurements are on humans and are basically subjective. The model has been tested by microelectrodes in the olfactory epithelium of frogs by Getchell (1974) and Gesteland et al. (1965). The nerve impulse rate was studied when both sterically related molecules and molecules with

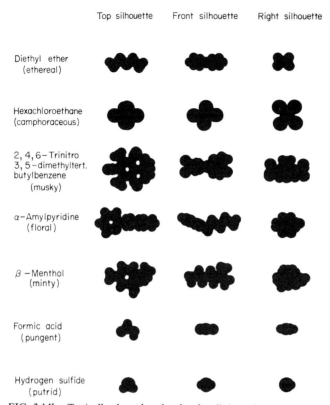

FIG. 3.14b   Typically shaped molecules that fit into the receptor sites.

comparable odor were blown across the epithelium. It was found that some receptors responded in the same manner to sterically related molecules while others did not. A similar diversity of response was found in the comparable odor test. There is clearly not a one-to-one odor–receptor correspondence, which makes the understanding of olfaction much more difficult.

PUNCTURE THEORY OF OLFACTION

The location of the reaction site between the stimulus molecule and the receptor is on the cilia. Whatever this reaction is, it must cause an influx of Na^+ ions and a K^+ efflux from the axon, as well as the motion of other ions, to produce an action potential. There are two viewpoints to the production of this action potential, which may be called the chemical and the physical.

In the chemical model, the stimulus molecule is assumed to attach to a receptor protein molecule. This temporary attachment causes either an electronic or a conformational change which in turn changes the Na^+ permeability of the axon. Some of these ideas will be briefly discussed.

The physical model, about to be described, starts with the assumption that there is a lipid (fatty substance) membrane which covers the end of the nerve axon. An odor molecule, after diffusing through this layer, temporarily leaves a hole between the molecules of the lipid layer. This hole, open for a very short time, permits K^+ ions to leak through. This decrease of K^+ ions in the axon permits some of the excess Na^+ ions, external to the axon, to enter the axon. This event triggers the action potential impulse which then is self-propagating along the axon (Chapter 1). The lipid layer under consideration is probably just a few molecules thick and is covered with a watery mucus solution.

Figure 3.13 shows that the difference in concentration for threshold stimulation between different molecules can be many orders of magnitude. The puncture theory of Davies proposes that while only one molecule of a strong odorant is necessary to create the hole, a cluster of several molecules of a weak odorant is required.

The justification for this cluster model of puncture was established by an auxiliary experiment. Davies and Taylor (1954) measured the lysis (dissolving) of red blood cells by saponin and similar hemolytic agents. The red blood cells (RBC) are covered with a thin lipid membrane. They measured the acceleration of the lysis when the RBCs were simultaneously exposed to odorants at concentrations of their respective olfactory thresholds. Figure 3.15 shows that β-ionone, the strongest odorant, requires a concentration of only 10^{-8} that of methanol. The reasoning, then, is that the penetration of the odorant through the membrane wall of the cell assists the hemolytic agent in its penetration, thereby accelerating hemolysis. It should be noted, however, that log–log plots are not very sensitive and can be deceptive.

In the development of the calculations for the puncture model, Davies makes the following definitions:

p = critical number of molecules required to be in one spot at the same time to make puncture hold of proper size.

x = mean number of molecules absorbed per square centimeter of surface.

c = the concentration of odorant molecules in air per cubic centimeter.

d = the thickness of the membrane.

K = absorption coefficient between air and lipid membrane.

n = surface sites of a receptor cell.

α = area in square centimeters of each n surface site.

N = the number of the n sites with p molecules.

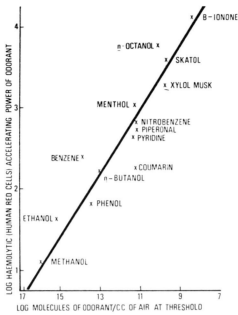

FIG. 3.15 Hemolytic acceleration in the presence of various odorants in concentrations at the threshold of olfactory detection. [Reprinted by permission from J. T. Davies and F. H. Taylor, *Nature* **174**, 693 (1954). Copyright © 1954 MacMillan Journals Limited.]

Figure 3.16 is a schematic of the membrane, where M are the odorant molecules.

The value of K can be independently determined by use of Eq. (3.4)

$$\Delta G^0 = 2.3 RT \log K \qquad (3.4')$$

where ΔG^0 is the total free energy change for an odorant molecule in the lipid layer compared to the molecule in air. As seen in Fig. 3.17, total free energy change is the sum of the changes from air to the aqueous-mucus

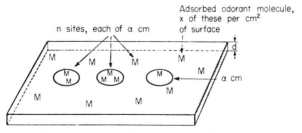

FIG. 3.16 Schematic of the wall of olfactory cell of thickness d containing n sensitive regions (or sites), each of an area of α cm^2. Odorant molecules M are adsorbed to a mean coverage of x molecules/cm^2. [From Davies (1973).]

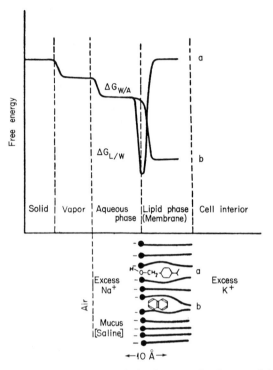

FIG. 3.17 Free energy changes for typical odorant molecules *a* and *b* passing from a solid, from which they evaporate, through various regions into the olfactory receptor membrane. *a* is a polar molecule and *b* a nonpolar one. [From Davies and Taylor (1957).]

layer and from there to the lipid layer. The free energy from air to the aqueous phase was calculated from the ratio of the odorant's solubility in water to its vapor pressure in air. The free energy from the water to the lipid phase was obtained from experiments by Davies and Taylor (1957) which measured the absorption coefficients of the odorant molecules on a water–lipid interface. They performed this experiment on 28 odorant molecules. Substitution of the sum of these two free energies into Eq. (3.4') yielded the separate values of K for each odorant.

The Langmuir (1916) isotherm[†] for the absorption of a monolayer of molecules on a membrane of thickness d (d = about 10 Å) can be written as

$$x/d = cK/(1 + cK) \qquad (3.6)$$

[†] The Langmuir isotherm is derived by a method analogous to that of Eq. (2.3). Using Langmuir's notation, let

μ = number of molecules striking unit surface in unit time.

It has been shown, Fig. 3.15, that for olfaction c is very small and therefore the second term in the denominator of the right-hand side of Eq. (3.6) is small compared to unit and

$$x = cdK \qquad (3.6')$$

is a valid approximation.

The area of the region where p molecules of odorant must cluster is α, and the average number of molecules in this region is αx. When this average number is equal to p or greater, a threshold detection event will occur. If the coverage of the surface is large and the number of events small, i.e., when $\alpha x \geqslant p$, the number of events N is given by the Poisson distribution (Vol. I, Appendix C, Section C.2),

$$N = ne^{-\alpha x}(\alpha x)^p/p! \qquad (3.7)$$

If we consider the occurrence of a single event, $N = 1$, then

$$ne^{-\alpha x}(\alpha x)^p = p!$$

and, upon taking the logarithm of both sides and substituting Eq. (3.6'), this yields

$$\ln c + \ln K + \ln \alpha d + \frac{\ln n}{p} - \frac{\alpha cdK}{p} + \frac{\ln p!}{p}$$

Experimental values show that the last term on the left-hand side is of the order of 10^{-3}, so it is neglected. Define c = olfactory threshold concentration (O.T.), i.e, the number of molecules of odorant/cm^3, for threshold discrimination, convert to base 10 logarithms ($\ln x = 2.3 \log x$), and this equation may be written as

$$\log \text{O.T.} + \log K = -\frac{\log n}{p} - \log \alpha d + \log \frac{p!}{p} \qquad (3.8)$$

α = the fractional rate at which this striking gas is absorbed (not greater than unit).

N_0 = number of absorption spots per unit surface area.

θ = fraction of surface that is bare.

θ_1 = fraction of surface covered, i.e., $\theta = 1 - \theta_1$

ν_1 = rate of evaporation of gas from surface if surface were completely covered.

Therefore, the rate of condensation = $\mu \alpha \theta$ and the rate of evaporation = $\nu_1 \theta_1$. At equilibrium, $\mu \alpha \theta = \nu_1 \theta_1$.

Upon substituting for θ, Langmuir obtained the amount of gas absorbed

$$\theta_1 = \mu \alpha/(\nu_1 + \mu \alpha)$$

This is called an absorption isotherm because both α and ν_1 are temperature dependent.

In the present application, μ is proportional to the concentration c and the forward rate constant is $\mu \alpha = K_1$. The back reaction rate constant is ν_1 and the amount of absorbed gas θ_1 per unit thickness of absorbed molecules is $\theta_1 = x/d$. With the usual definition that $K = K_1/K_2$ Eq. (3.6) can be written.

The unknowns n and αd may be determined by using the data of Fig. 3.15 for O.T. of the strongest and weakest odorants, the experimental values of K, and reasonable choices for p.

For β-ionone, let $p = 1$ and O.T. $= 1.6 \times 10^8$ molecules/cm^3 (log O.T. $= 8.2$), and the experimental value of the rate constant of this odorant is log $K = 8.35$. Substitution of these values into Eq. (3.8) yields

$$8.2 + 8.35 = -\log n - \log \alpha d$$

For a weak odorant take $p = 24$ (for calculational convenience since $\log 24!/24 = 1$). Experimentally, for weak odorants, the approximate values are log O.T. $= 18$ and log $K = 4$. Substituting these values into Eq. (3.8) yields

$$18 + 4 = -\tfrac{1}{24}\log n - \log \alpha d + 1$$

Solution of these equations simultaneously yields the result $\log \alpha d = -21.2$ and $\log n = 4.65$.

If $\log n = 4.65$, then $n = 44{,}000$, the number of surface sites on each receptor cell. The area of each of these sites is α, and d, the thickness of the absorbed monolayer, can be assumed to be about 10 Å. Hence, since $\log \alpha d = -21.2$, α is 64×10^{-16} cm^2 or 64 Å^2, which is not an unreasonable value. The product αn represents the active area of each olfactory cell, and from these values it is 2.8×10^{-10} cm^2. However, physiological measurements of the cell give values from 10^{-8} to 10^{-6} cm^2. It must be noted from Fig. 3.3 that only the very fine cilia, or hairs, are exposed to the odorant, not the entire cell. It is seen in this figure that it is not unreasonable that the active area is several orders of magnitude smaller than the cell area.

From the definition of p, the size of the cluster required to make a proper-size puncture, i.e., one open sufficiently long enough to permit an adequate transport of ions (Davies, 1965), a reciprocal quantity $1/p$ may be defined as a "puncture factor." This might be expected to be linearly proportional to the cross-sectional area A_0 that is required to make the proper size puncture. With the previous determination of the constants n, αd, and K, and with the estimate of p from the above linear relationship, theoretical values of O.T. can be calculated from Eq. (3.8). Such theoretical values are compared with experimental measurements for a variety of odorants in Fig. 3.18, in which open circles are normal alcohols, triangles are normal hydrocarbons, and closed circles are other organic compounds. Adjustment of the slope to give the best fit to the data yields a value of $A_0 = 47$ Å^2, in close agreement with the previously determined value of 64 Å^2. It should be noted that these calculations do not exclude the steric model which may involve a "lock and key" effect.

Davies (1965) modified his model somewhat. Since there are two critical

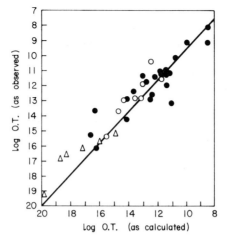

FIG. 3.18 Comparison of measured olfactory thresholds for humans compared to those calculated from Eq. (3.10) for a variety of compounds. O, normal alcohols; ▲, normal hydrocarbons; ●, other organic compounds. [From Davies and Taylor (1959).]

parameters, hole size and time it remains open, he has further postulated that different odor receptors have membranes of slightly different physical characteristics. Readers interested in this topic should not omit this paper.

CHEMICAL STUDIES OF RECEPTOR MOLECULES

Although the electrophysiology, stereochemistry, and physical theory approaches all have merit and may all prove to be a part of the grand scheme of olfaction, these techniques seem to be approaching a limit. That limit is the state at which none can be proven right or wrong, or are even able to proceed further in the elucidation of the mechanism. More information is clearly needed and modern chemical techniques may be able to provide it. We will very briefly introduce the reader to some of the literature of this approach, but it should be recognized that any new technique has taken many years and the contributions of many researchers to develop. The omission of the work of important contributors is not an oversight, but a comprehensive review is not the aim of this simple introduction. The reader, looking up one or two of the references, will quickly be led into the literature of this exciting field.

Ash (1968) originally demonstrated that the olfactory epithelium of a rabbit, when homogenized, contained a fraction which exhibited a decrease in its ultraviolet spectrum at 267 nm when exposed to moderate amounts of the odorant linalool. He later showed (1969) that this decrease requires the presence of ascorbic acid (vitamin C) as a cofactor. Ascorbic acid is bound to proteins in the olfactory epithelium, but when an odorant is attached, the ascorbic acid is released and oxidized. His studies indicate that this release and oxidation is the origin of the decrease of the 267-nm spectral absorbance. He suggested that the odorant release of the ascorbic acid may manifest itself as a conformal change in the receptor protein which can trigger the action potential. However, Cagan and Zeiger (1978) disagree with the findings of Ash.

Getchell and Gesteland (1972) presented evidence that the olfactory receptor of a frog is a protein. They did this by recording from olfactory nerve axons while subjecting the epithelium to the odorant ethyl n-butyrate. The response could be neutralized by first exposing the epithelium to N-ethylmaleimide. This latter chemical is one of a class of group-specific protein reagents. These reagents are in wide use in the study of pure enzymes because when such a reagent is added, the activity of the enzyme is reduced or abolished. From these latter extensive studies, a correlation may be made with one or more amino acid residues on the enzyme which are responsible for its activity. If the nose of the frog was first saturated with the odorant ethyl n-butyrate, the action of the reagent was blocked because a simple wash restored the activity of the epithelium. Now, ethyl n-butyrate is a small molecule, and this experiment therefore showed that its attachment at the active site of a large protein molecule could block the attachment of the neutralizing reagent.

Fesenko et al. (1977) have reported that they have successfully separated the active molecular fraction from a frog's olfactory epithelium and have partly characterized it. They accomplished this by first homogenizing the epithelium and then centrifuging it. An artificial membrane was formed from brain lipids, by what is now a standard technique, in the presence of the clear supernatant. When this membrane was exposed to odorants, its permeability to both Na^+ and Ca^{2+} ions was increased dramatically. Gel filtration showed the active ingredient to be a nucleoprotein with a molecular weight of no less than 100,000. See Cagan and Zeiger (1978).

Not to be neglected is the work of Laffort and Dravnicks (1973), who are developing a model of odorous compounds based on physicochemical parameters. Also, Laffort (1969) has shown that it is possible to calculate odor thresholds of 50 compounds of quite different types, carrying different functional groups, from four physicochemical parameters. This type of work clearly anticipates the next step of understanding, which is how and why certain odorants become attached to the active protein molecules.

These early experiments indicate that techniques will be developed to study the receptor protein in vitro and thereby determine its characteristics. One can foresee the development of standardized odor concentration measuring instruments by the incorporation of these molecules into artificial membranes.

REFERENCES

Adrian, E. D. (1950). The electrical activity of the mammalian olfactory bulb, *Electroenceph. Clin. Neurophysiol.* **2**, 377.

Amoore, J. E. (1964). Current status of the steric theory of odor, *Ann. N.Y. Acad. Sci.* **116**, 457.

Amoore, J. E., Johnston, J. W., Jr., and Rubin, M. (1964). The stereochemical theory of odor, *Sci. Amer.* **210** (Feb.), 42.

Ash, K. O. (1968). Chemical sensing: an approach to biological molecular mechanisms using difference spectroscopy, *Science* **162**, 452.

Ash, K. O. (1969). Ascorbic acid: cofactor in rabbit olfactory preparation. *Science* **165**, 901.

Cagan, R. H., and Zeiger, W. N. (1978). Biochemical studies of olfaction: Binding specificity of radioactively labeled stimuli to an isolated olfactory preparation from rainbow trout (*Salmo gairdneri*), *Proc. Nat. Acad. Sci. U.S.* **75**, 4679.

Cheesman, C. H., and Mayne, S. (1953). The influence of adaptation on absolute threshold concentrations for olfactory stimulus, *Q. J. Exp. Psychol.* **5**, 22.

Davies, J. T. (1962). The mechanism of olfaction, *in* "Biological Receptor Mechanisms" (*Symp. Soc. Exp. Biol.*), Vol. XVI. Cambridge Univ. Press, London and New York.

Davies, J. T. (1965). A theory of the equality of odours, *J. Theoret. Biol.* **8**, 1.

Davies, J. T. (1973). Olfactory theories, *in* "Handbook of Sensory Physiology," Vol. IV. Springer-Verlag, Berlin and New York.

Davies, J. T., and Taylor, F. H. (1954). A model system for the olfactory membrane, *Nature* (*London*) **174**, 693.

Davies, J. T., and Taylor, F. H. (1957). Molecular shape, size and adsorption in olfaction, *in* "Solid/Liquid Interface and Cell/Water Interface" (*Proc. Int. Congr. Surf. Activity, 2nd*), Vol. IV. Butterworths, London.

Davies, J. T., and Taylor, F. H. (1959). The role of adsorption and molecular morphology in olfaction: the calculation of olfactory thresholds, *Biol. Bull. Woods Hole* **177**, 222.

DeLorenzo, A. J. D. (1963). Studies on the ultrastructure and histophysiology of cell membranes, nerve fibers and synaptic junctions in chemoreceptors, *in* "Olfaction and Taste" (Y. Zotterman, ed.). Pergamon, Oxford.

Fesenko, E. E., Novoselov, V. I., Pervukhin, G. Ya., and Fesenko, N. K. (1977). Molecular mechanisms of odor sensing. II. Studies of fractions from olfactory tissue scrapings capable of sensitizing artificial lipid membrane to action of odorants, *Biochem. Biophys. Acta* **466**, 347.

Gesteland, R. C., Lettvin, J. Y., Pitts, W. H., and Rojas, A. (1963). Odor specificities of the frog's olfactory receptors, *in* "Olfaction and Taste" (Y. Zotterman, ed.). Pergamon, Oxford.

Gesteland, R. C., Lettvin, J. Y., and Pitts, W. H. (1965). Chemical transmission in the nose of the frog, *J. Physiol.* **181**, 525.

Getchell, T. V. (1974). Unitary responses in frog olfactory epithelium to sterically related molecules at low concentrations, *J. Gen. Physiol.* **69**, 241.

Getchell, M. V., and Gesteland, R. C. (1972). The chemistry of olfactory reception: stimulus-specific protection from sulfhydryl reagent inhibition, *Proc. Nat. Acad. Sci. U.S.* **69**, 1494.

Guyton, A. (1971). "A Textbook of Medical Physiology." Saunders, Philadelphia.

Johnston, J. W. (1963). An application of the steric odor theory, *Georgetown Med. Bull.* **17**, 40.

Laffort, P. (1969). A linear relationship between olfactory effectiveness and identified molecular characteristics extended to fifty pure substances, *in* "Olfaction and Taste" (C. Pfaffmann, ed.). Rockefeller Univ. Press, New York.

Laffort, P., and Dravnicks, A. (1973). An approach to a physico-chemical model of olfactory stimulation in vertebrates by single compounds, *J. Theor. Biol.* **38**, 335.

Langmuir, I. (1916). The constitution and fundamental properties of solids and liquids, *J. Am. Chem. Soc.* **38**, 2221.

Moncrieff, R. W. (1967). "The Chemical Senses." CRC Press, Cleveland, Ohio.

Ottoson, D. (1956). Analysis of the electrical activity of the olfactory epithelium, *Acta Physiol. Scand. Suppl. 122* **35**, 1.

Shibuya, T. (1965). Dissociation of the olfactory neural response and mucosa potential, *Science* **143**, 1338.

Tucker, D. (1963). Physical variables in the olfactory stimulation process, *J. Gen. Physiol.* **46**, 453.

Zotterman, Y., ed. (1963). "Olfaction and Taste." Pergamon, Oxford.

CHAPTER 4

Cutaneous Sensation

INTRODUCTION

Investigators in the past have located a variety of types of nerve endings in the skin, and these are generally named after them. One encounters such items as Meissner corpuscles, Merkel's disks, Krause end bulbs, Ruffini endings, and Pacinian corpuscles. See reviews by Sinclair (1967) or Andres and von Düring (1973) for illustrations. The reader should remember that just because an object has a name does not mean that it is understood. In fact, of the above-named set, only the Pacinian corpuscle is reasonably understood, and this will be described in detail later. Small endings of cells whose role is suspected are simply too small to insert microprobes for study. Only in some of the larger corpuscles have electrophysiological studies been made, but even though their role is identified, their mechanism is still beyond experimental approach. The action of biological transducers, systems that convert mechanical to electrical energy, has been observed, but the cellular process of this is still unknown.

There are two basic types of skin, glabrous (without hair) and hairy. The glabrous palm of the hand contains (1) free nerve endings, (2) branched fibers with expanded-tip endings, and (3) encapsulated endings. The hairy skin of the back of the hand contains (1) free nerve endings and (2) endings associated with the hair follicle. These latter consist of (a) free endings, (b) endings with expanded tips, and (c) occasional encapsulation of the above structures (Miller *et al.*, 1960). Thus, although hairy skin contains the same type of nerve endings found in glabrous skin, these have a different pattern of arrangement.

In addition to the smallness of the receptors, their identification is further complicated by two factors. One is that many receptors are polymodal, they are sensitive to both mechanical and temperature changes. The other is that many cells are connected by the same axon which then passes through one or more synapses to the central nervous system. Thus, a signal from a receptor to the brain does not necessarily have a one-to-one correspondence, and a measurement past the first synapse is not necessarily a record of the electrical response of the receptor to the stimulus.

The first electrical signal in a nerve from a cutaneous receptor was recorded by Adrian and Zotterman (1926). Since that time, an enormous amount has been published as scientists have tried to map out the pathways of sensation and response through the neural network and brain. A large part of the mapping has been successful but, because of synapses and cross-linking of nerve fibers, much of the information remains simply a map. The detailed chemistry at the various receptors remains unknown and, until it is known, treatment for afflictions such as pain or itching remains an art rather than a science. The development of microchemical techniques to complement the electrophysiological measurements remains as one of the great physiological advances yet to be made.

It will be seen in this chapter that when sound experimental data are available, scientists are able to make reasonable interpretations. It will also be seen that the beginnings of the nerve codes are being elucidated, specifically, from a series of action potentials how can the brain tell what is happening at a certain point in the body? The code is observed, but the method of deciphering has not yet been determined. With the present availability of high speed computers to simulate the varieties of signals, it would appear that, with more guidance to the electrophysiologists from the cryptographers of the type of data they need, the code could soon be broken.

In this chapter, we will describe some of the measurements and show the physical science that has been applied to interpret the behavior of the Pacinian corpuscle, a displacement velocity sensor in the body.

In the next section, we will describe for illustration three of the many types of mechanoreceptors.

MORPHOLOGY OF EXAMPLE MECHANORECEPTORS

There are many types of free endings of nerve fibers in the skin and many of these surround hair follicles. A single axon may also be afferent from several hairs, and the sensory endings may coil around the base of the hair or be parallel to it. Figure 4.1 shows a drawing of receptors around a hair in human skin. The fibers will be disturbed when the hair is distorted. It is probably a rapidly adapting (RA) type of mechanoreceptor which is responsive only to velocity both in initial distortion and in the return of the hair to equilibrium.

A more complicated arrangement is in the Meissner corpuscle in glabrous skin. A schematic drawing from an electromicrograph is shown in

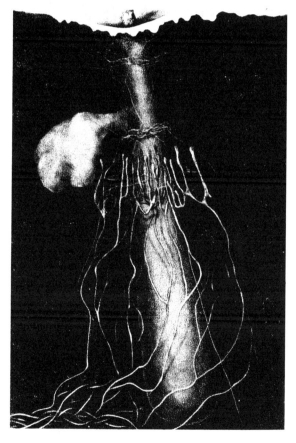

FIG. 4.1 Drawing from an electronmicrograph of the disposition of nerves around a small hair follicle. [From Montagna and Parakkal (1974).]

Fig. 4.2. This complex is also probably a rapidly adapting mechanoreceptor and responds to high frequency mechanical vibration. It is seen that
there are small fibrils attached both externally to the epidermis and
internally to the cells of the corpuscle. A small disturbance of the skin in
any direction will cause a transducer action within the corpuscle, probably
chemical, and thereby generate an action potential in the axon.

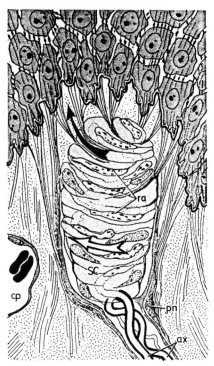

FIG. 4.2 Schematic of a Meissner corpuscle. The black arrow represents a direction of
fibril motion which acts on the corpuscle and the white arrow is a consecutive movement
which would relieve the strain. This sequence could explain the rapid adaptation of the
receptor and its corresponding response to high frequency disturbance. ra, receptor axon; sc,
Schwann cells; pn, perineural sheath; cp, capillary. [From Andres and von Düring (1973).]

The Pacinian corpuscle has an interesting structure, something like an
onion. A microphotograph is shown in Fig. 4.3a and a cross section in Fig.
4.3b. The inner core is a nonmyelinated axon which becomes myelinated
as it leaves the corpuscle. The successive layers of membrane are separated
by a fluid similar in viscosity to water. It responds to rapid changes in
distortion but only slightly to static distortion. We will discuss the mechanical basis of this corpuscle's design later in this chapter.

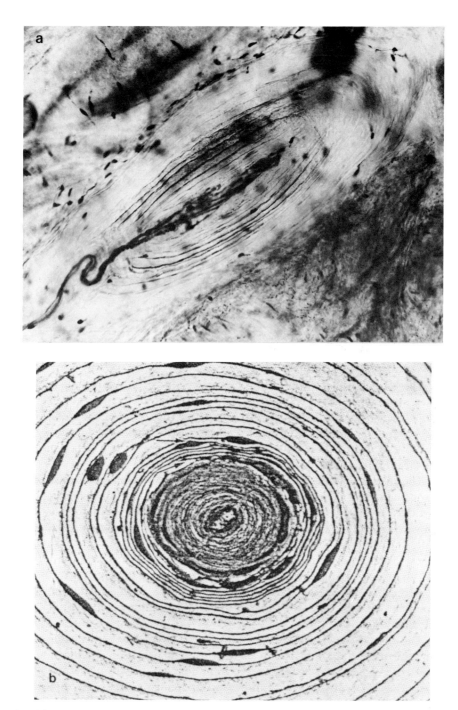

FIG. 4.3   Pacinian corpuscle. (a) Cross section parallel to axon. [From Miller *et al.* (1960).]
(b) Cross section perpendicular to axon. [From Andres and von Düring (1973).]

ROUTE OF SOMATOSENSORY SIGNALS

There are a variety of types of fibers from cutaneous sensors, some large and myelinated and others small and nonmyelinated. The former carry the signal rapidly, while others transmit more slowly. Table 4.1 shows the modern classification of these fibers and some facts about them.

TABLE 4.1

Afferent fibers and cutaneous receptor type[a]

Fiber classification	Aα		Aδ	C
	Group I	Group II	Group III	Group IV
Diameter	20–12 μm	11–6 μm	5–1 μm	1.5–0.3 μm
Myelinated	yes	yes	yes	no
Conduction velocity	120–72 m/sec	72–30 m/sec	30–4 m/sec	2–0.4 m/sec
Receptor	touch	touch	pain, heat, cold, pressure	pain, polymodal

[a] Based on data collected by Burgess and Perl (1973).

These fibers do not carry the signals directly to the brain, but instead they synapse in neuronal pools. There are two transmission systems for somatosensory information in the central nervous system, the *dorsal column* and the *spinothalmic* systems. The axons from the sense detectors first synapse in one of these two systems before the information is transmitted further (Lynn, 1975).

The dorsal column systems contain the large myelinated axons. Such axons can transmit information rapidly and carry additional information such as precise location and fine sensitivity. The spinothalmic system consists of smaller axons which carry more general information at a slower rate, such as heat and cold, pain, tickle and itch, and sexual sensations. These different systems represent another example of an organism reducing the energy requirement for growing and maintaining its parts. Those required for survival have no effort spared, while others, related more to comfort than survival, are relegated to a lower position in the evolutionary hierarchy.

The sensory receptors are distributed in a defined pattern on the skin, and this pattern has almost an exact correspondence at precise locations in the brain, except that a single nerve from a receptor does not terminate at a spot in the brain. This one-to-one correspondence has been studied in detail with subjects under local anesthesia with part of the skull removed.

An older map, but still essentially valid, is seen in Fig. 4.4. If the different points indicated are lightly touched, sensation is felt by the subject in the corresponding parts of his skin. Note that the sensation is always on the side of the body *opposite* to the side of the brain that is touched. An even more detailed map of this area is shown in a cross section in Fig. 4.5, in which it is seen that the face, particularly the lips, occupies a disproportionately large surface area of the cortex. In a converse experiment with primates, an electrode is inserted in the cortex, the other grounded to the scalp, and different parts of the skin touched until a potential is observed. The mapping results are essentially the same as the experiments on man. What has also been learned is that the cerebral cortex has six separate layers of neurons and that the incoming sensory stimuli excite the neurons in layers 3 and 4 first. The signal then spreads outward and inward until all six layers are excited, with some of the excitation spreading laterally. The reason for this is not known.

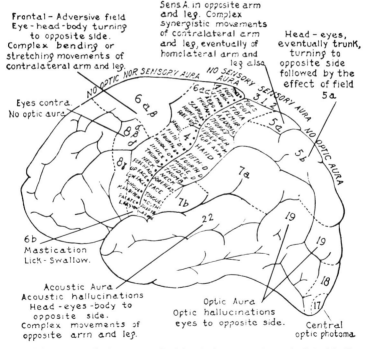

FIG. 4.4 Diagram of the human cerebral hemisphere seen from the left side illustrating areas whose electrical stimulation is followed by movement or sensation. [From Herrick (1931).]

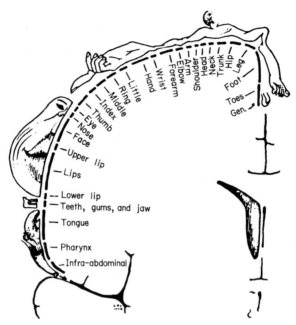

FIG. 4.5 Cross section of the brain with the length of the black bars indicating the relative proportion of each part of the brain to the position of stimulation. The figures are drawn with appropriate exaggeration. [Reprinted with permission from W. Penfield and T. Rasmussen, "The Cerebral Cortex of Man," Macmillan, New York, 1950. Copyright © 1950 by Macmillan Publishing Co., Inc., renewed 1978 by Theodore Rasmussen.]

SINGLE-FIBER RECORDING FROM HUMANS

A technique has been developed for the insertion of tungsten microelectrodes through the skin into the nerve fibers of conscious human subjects. This has permitted the simultaneous, or sequential, measurement of the magnitude of a signal from a touch receptor and its comparison to the perceived, or psychophysical, sensation. When the electrode is in an appropriate fiber and the corresponding receptor area in the hand is located, the relative magnitudes of the electrical response (ratio of frequency of impulses to the maximum frequency obtainable) can be measured for different stimulus intensities. In the present case, skin indentation amplitude was the stimulus. At the same time, the subject is asked if the amplitude is greater or lesser than the previous stimulus. This type of psychophysical determination is called a "two alternative forced choice" (2 AFC) experiment. Comparative results for a rapidly adjusting mechanoreceptor on the thumb of a subject are shown in Fig. 4.6, where the ordinate for the psychophysical measurement indicates the percentage of correct

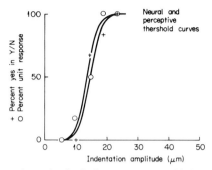

FIG. 4.6 Comparison of psychophysical measurement with impulse frequency from a nerve axon when the human hand is stimulated by alternating indentation of the skin. O, nerve impulse rate; +, perception threshold. [From Vallbo and Johansson (1976).]

responses. The open circles indicate the electrical responses. Although this good agreement is typical for the fingers, similar data for the palm of the hand, for which the threshold is higher, do not compare as favorably. Other experiments of this type have been performed on human subjects by Vallbo and Hagbarth (1968) and Van Hees and Gybels (1972). Clearly, much work is yet to be done in this field.

THERMORECEPTORS

The identification of thermoreceptors in the skin is not complete. It is known, however, that they are not one of the readily identifiable encapsulated complexes of the skin. The receptors are different for cold and warm sensations, that is, there are cold and warm receptors. This has been determined by electrophysiological measurements in a nerve bundle in an arm or leg of a vertebrate mammal. There are two types of cold and warm receptors, one is sensitive to steady-state temperature, while the other is sensitive to the rate of change of temperature. Examples of receptor densities in human skin are given in Table 4.2.

TABLE 4.2

Number of cold and warm spots per cm^2 in human skin^a

	Cold spots	Warm spots
Nose	8	1
Face	8–9	1.7
Chest	9–10	0.3
Forearm	6–7	0.3–0.4
Back of hand	7	0.5
Palm of hand	1–5	0.4

^{a}Data from Hensel (1973).

It is believed that there is a steady-state diffusion of ions across the membrane of the thermoreceptor and that the forward and back rate constants have different temperature dependences. Thus, a change in temperature will alter the ion concentration in the receptor and generate an action potential.

Studies of the electrical signals in the axons of steady-state thermoreceptors have been made by Iggo (1969) on cats, rats, and monkeys. These receptors provide continuous information on their temperature by electrical impulses, a steady rate at a given temperature in the cat, but bursts of impulses in the monkey. The number of impulses per second is a function of temperature, and Fig. 4.7 shows the impulse rate at different temperatures in rat axons for both warm and cold receptors. The skin temperature of the rat is normally about 33°C, at which temperature the warm axons are inactive and the cold axons are delivering impulses at a low rate. The comparison of thermoreceptor responses from different species is not strictly valid. Earlier work on both warm and cold receptors in a cat's tongue has shown that there is overlap of the two signals at body temperature and that the firing rate never goes to zero. The cold receptors fire at a steady rate, while the warm receptors fire in bursts, and thus the brain has secondary cues to distinguish between steady cold and steady warmth.

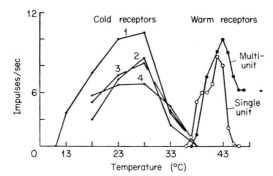

FIG. 4.7 A comparison of cold and warm thermoreceptor responses to steady-state skin temperatures. [From Iggo (1969).]

The rate of change of temperature of the skin is very sensitive, as experienced in briefly touching a piece of metal or feeling a sudden cool breeze or hot breath on the skin. Experiments have shown that such temperature change effects in the palm of a monkey and a human are similar; since experiments on electrical impulses in axons require an operation on the arm, monkeys are generally used. The experiments of rate-of-change cold receptors in monkeys performed by Darian-Smith et al. (1973) will be described.

A hollow silver probe was used for the temperature contact on the skin. The tip of the probe was about 1 mm in diameter, and the probe was hollow to admit an aqueous coolant. The temperature change of the tip of the probe, exponential in time, fell to $1/e$ of its initial value in about $\frac{1}{4}$ sec and achieved the new value within $\frac{1}{2}$ sec. Examination of about 500 axons, in several monkeys, showed that each axon responded to the temperature change of a spot on the palm, most spots being less than 1 mm in diameter.

In the experiment, a microelectrode was inserted into an axon and the particular spot associated with that axon located on the palm and the cold probe placed against it. The equilibrium temperature of the palm, called the T-base, was 34°C. The temperature of the probe would then be lowered a few degrees, called a T-step, and the signals in the axon recorded as a function of time. Figure 4.8 shows a typical result from a single axon. Sufficient time between T-steps was allowed for the steady-state rate at 34°C to be reestablished. The upper curve in this figure is a profile of the temperature–time curve of the probe with an arbitrary ordinate.

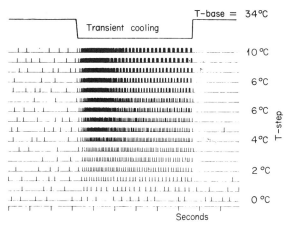

FIG. 4.8 Responses of a cold nerve fiber elicited by cooling pulses graded in intensity. The T-base is 34°C and the T-steps are shown at the right. The upper trace is a profile in time of the application of the T-step. [From Darian-Smith et al. (1973).]

It is seen in Fig. 4.8 that there is an intense burst of impulses shortly after the temperature drop, after which the number of impulses per second decreases. It is also seen that the greater the temperature drop, the greater the intensity of the impulses.

How does the brain sort out this information? Darian-Smith et al. were able to show the code. The impulse rate of Fig. 4.8 at each temperature was normalized to an arbitrary scale of 100 at 4 sec. The resulting profiles of impulses versus time for *T*-steps of 2, 4, 6, and 8°C are normalized to the same height in the histograms on the diagonal of Fig. 4.9, i.e., the single histograms. Each normalized histogram is then successively superposed on those in the *T*-step rows above it by a vertical displacement. Very good correlation is seen. The histogram code of impulse rate from this particular thermoreceptor through its particular axon is quite different from others. To illustrate this striking difference, the histogram codes from two other thermoreceptors are shown in the inset of Fig. 4.9. Thus, the brain is supplied with information from which it can tell both the location and the amount of the temperature change. How the brain decodes this information must be left for future investigation (see pp. 32–34).

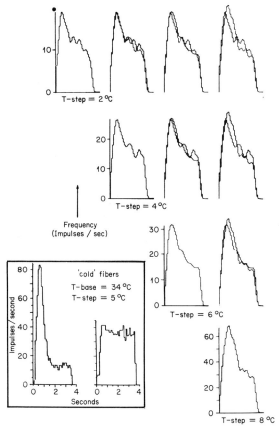

FIG. 4.9   Normalized histograms (diagonal figures) of the discharge data of Fig. 4.8. The rows are duplications of the first member of a row. Superposition is obtained by moving members of the next lower row upward. The inset shows the pattern from two other fibers. [From Darian-Smith *et al.* (1973).]

NOCICEPTORS

Above a certain sensitivity threshold, mechanical or thermal stimulation of the skin becomes painful. Special units with both myelinated and nonmyelinated axons exist in the skin, called *nociceptors* or pain receptors. Although the receptors have not been positively identified, electrical probes placed in the afferent axons have indicated their existence. These axons were located by trial and error exploration of nerve bundles. Some were identified as not giving rise to discharges unless the skin were stimulated to a damaging intensity. In general, three classes of nociceptors have been identified, although there are probably subgroups: mechanical, thermal, and polymodal nociceptors which respond to both types of stimuli. For example, a typical thermal nociceptor will not generate impulses until the temperature exceeds 42°C, while another type will not generate impulses above 10°C. Figure 4.10 shows the rate of impulse generation of a high temperature thermal nociceptor of a cat when a metal rod at the indicated temperatures is placed against the skin. The particular unit illustrated here was very insensitive to mechanical stimulation.

FIG. 4.10 Thermal nociceptor axon response to the application to the skin surface of a metal rod at the temperatures indicated on the left. [From Iggo (1977).]

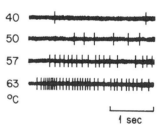

Although there are nociceptors which are sensitive only to mechanical stimulation, quantitative measurements are much more difficult. The skin can be pricked, pinched, chafed, or cut to a damaging level, and recordings of the signals in the afferent nerve fibers made, but how does one calibrate relative damaging stimuli? This matter has not yet been resolved.

It is known from experience that there is a "first," or initial, pain followed by a lasting or "second pain." Because of the speed of reaction to the first pain, it is expected that this must be conducted by myelinated nerve fibers in the dorsal column. The slower second pain can travel by means of the morphologically simpler nonmyelinated fibers which have a lower energy of creation and repair. Experiments have confirmed this. The majority of pain-conducting fibers are among the smallest of all nerve fibers with conduction rates below 2 m/sec (Burgess and Perl, 1967).

CUTANEOUS HYPERALGESIA

Increased pain sensitivity of the skin, called *cutaneous hyperalgesia*, is a phenomenon experienced by all. It is the tenderness of the area surrounding an injury. It also happens with only minor injury, such as sunburn or scratching. A characteristic of the skin is to become reddened and hot because of dilation of blood vessels in the local region (Lynn, 1977).

Many studies have been made which sought the origin of this phenomenon and, although it is not yet understood, significant clues have been obtained. Nonmyelinated polymodal (simultaneous conductors of both thermal and mechanical stimuli) fibers have been studied from such regions. It has been learned that sensitization causes both a lowering of the threshold and an increase of signal frequency with temperature. An example of sensitization developed in 3 min is shown in Fig. 4.11 from studies

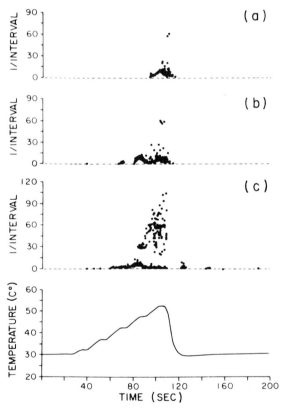

FIG. 4.11   Sensitization of C-fiber nociceptor. Abcissa is discharge rate in afferent fiber upon heating a monkey's skin in the pattern of the lower part of the figure with a 3-mm^2 thermode. (a) is first heating and (b) and (c) follow after 3-min rest between each heating. [From Kumazawa and Perl (1977).]

by Kumazawa and Perl (1977). The lower curve shows the pattern of applying noxious heat to a monkey's skin. The upper part of the figure is a plot of discharge rate recorded in a fiber leading from the nociceptor during the heating. The uppermost plot was obtained during the first heating. Following a rest of 3 min the heating pattern of the lower curve was repeated. The results are shown in the middle figure. There is clearly an increased signal rate at all temperatures. The increased rate beginning at 40°C is the new background or burning sensation produced by the injury of the first heating. In (c) the sensitivity begins at 35°C.

Careful studies of extracts of hyperalgesic areas have suggested several possible biochemical agents as the cause of hyperalgesic behavior, and injection in an area of uninjured skin of some of these has produced similar phenomena. Although the numbers of possible chemical candidates have been reduced, no firm conclusion has yet been reached nor is it known why these chemicals can lower the impulse threshold and increase the firing rate.

In addition to the lack of specific knowledge of the chemical mechanism, the mechanism of initiation is not known. That is, if the area of injury is first locally anesthetized, no tenderness appears in the surrounding areas until the effects of the anesthetic wear off from the injured area. This strongly suggests that the local receptors themselves do not signal the production of the pain-producing agent when they are injured, but instead the production is directed by some region in the central nervous system only when it receives a pain signal (King et al., 1976).

MECHANORECEPTORS

The sense of touch requires a receptor sensitive to mechanical distortion. This sensitivity, however, must be able to signal the difference to the brain between static distortion and distortion at different velocities. Once having sent its signal, if the distortion is not damaging to the organism, the signal should fade away so as not to clog the information channels with needless duplication. In addition, it should be adapted to its location in the body. For example, a receptor which signals the amount of inflation of the lungs is also near the heart and should therefore either not be stimulated by heart or pulsating arterial movements or else be able to send its meaningful signal between beats.

There are a variety of mechanoreceptors in skin with the general classification of rapidly adapting, RA, and slowly adapting, SA, both terms referring to their sensitivity to velocity of change, with the addition that the slowly adapting is also sensitive to static displacement. The organization of the mechanoreceptors in the skin depends, to a large degree, on whether the skin is hairy or glabrous. In hairy skin, there are mechanoreceptors in each hair follicle whose nerves discharge impulses only during displacement of the hair and when the hair is returning to its original position. For a survey of these receptors, see Burgess and Perl (1973).

One type of RA mechanoreceptor is the Meissner corpuscle. This seems to have small collagen fibers attached directly to the axon so that they pull when displaced and cause an action potential in the axon (see p. 72). An experiment performed by Dickhaus et al. (1976) on the afferent nerve of a Meissner corpuscle from the sole of a cat's foot illustrates the type of signal generated. They arranged a device with which the skin deformation could be controlled in amplitude, velocity, and duration. A summary of their typical results is shown in Fig. 4.12. In this figure, the mean discharge interval (the reciprocal of the discharge frequency) is plotted against the velocity of deformation of a constant amplitude. The four insets in the upper part show the time course of the discharge pulses, while the lower part of each shows the velocity of displacement on an arbitrary scale. Two things are evident in the example insets: (1) the discharge takes place only during the displacement and (2) the rate of discharge depends on the rate of displacement. A third thing also evident is that the rate of discharge

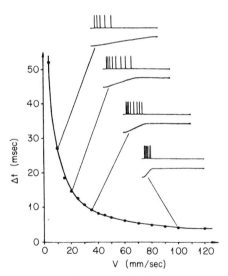

FIG. 4.12 Relationship in a single rapidly adapting mechanoreceptor between the mean discharge interval $\overline{\Delta t}$ and the velocity of the skin displacement V. Each point is the mean of all the time intervals between successive spikes of a discharge averaged from 20 trials. [From Dickhaus et al. (1976).]

decreases with time even though the velocity of displacement is constant. Therefore, only the first two or three signals to the brain can be used as accurate information. The average discharge interval was used to obtain the points of Fig. 4.12. Possibly, the brain does not need precision data and can also employ an average value over the time course of the displacement change.

TRANSDUCER EFFECTS IN THE PACINIAN CORPUSCLE

The Pacinian corpuscle, described earlier, is a system that is sensitive to touch. It consists of a nonmyelinated nerve ending surrounded by many layers of membrane (lamellae), similar to the layers of an onion, with a fluid between the membrane layers. A cross section of the central region of a corpuscle is shown in Fig. 4.3b. The corpuscle itself is more ellipsoidal than spherical in shape. The nonmyelinated nerve core becomes myelinated as it leaves. The results of a series of experiments in which strain is converted to voltage (*transducer effect*) by Loewenstein and associates will now be described.

Figure 4.13 illustrates the general experimental arrangement for compression, dissection, and electrical measurements. The myelin sheath of the axon is drawn in black and only the first two nodes of Ranvier are shown. The inner core is the nonmyelinated nerve ending enclosed by a thin sheath of lamellae, with an overall thickness of about 10 μm. This experimental arrangement permitted brief deflections of the lamellae, by means of the glass rod, of about 1-msec duration with variable amplitude up to 45

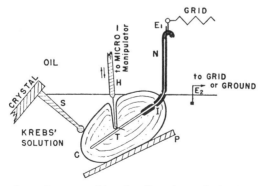

FIG. 4.13 Experimental setup used in microdissection and stimulation. C, corpuscle; T, nonmyelinated nerve ending; I, first Ranvier node (myelin drawn in heavy black with only first two nodes shown); S, glass stylus that transmits vibrations of the piezoelectric crystal; H, steel hook. [From Loewenstein and Rothkamp (1958).]

μm. The outer layers of the lamellae could also be removed and just the nonmyelinated nerve ending studied.

With the oscilloscope connected to the myelinated axon, where it is exposed at a node of Ranvier, it was found that the expected all-or-nothing potential rule was followed. That is, no signal was passed by the myelinated axon until a threshold mechanical deflection of the corpuscle was achieved. By a series of carefully placed probes, using an intact corpuscle, the investigators found that the nonmyelinated ending produced a potential if stimulated even though the potential was too low to induce the all-or-nothing propagation of the impulse through the myelinated axon. This is called the *generator potential* and is shown in Fig. 4.14. At 100% generator potential, sufficient voltage is produced for the myelinated axon to propagate the impulse. The lamellae of the corpuscle were then removed and incisions made in the protective sheath about the nonmyelinated ending. It was found that the magnitude of generator potential for a given stimulus strength at an exposed region was dependent upon the area of the exposed region, Fig. 4.15. That is, the magnitude of the generator potential seems to be localized and confined to axon regions mechanically activated, and these regions are independent of each other. If two regions are exposed, contacts placed on them and stimulated simultaneously, the currents add. This is seen in Fig. 4.15. The conclusion is that the total generator current is an integral of the local currents of spatially distributed membrane units.

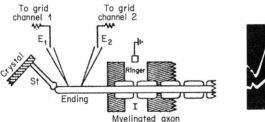 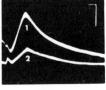

FIG. 4.14 Local activity confined to a small membrane portion of ending. Two microelectrodes connected to separate amplifier channels are placed about 350 μm apart in contact with the nonmyelinated ending of decapsulated corpuscle. E_1 is placed at a distance of approximately 20 μm from the stimulating stylus St. The first Ranvier node I is grounded and insulated from the ending. The upward deflection of the beam means "ending negative." Beams 1 and 2 give the potentials in response to a single mechanical stimulus, as recorded by electrodes E_1 and E_2, respectively. Note the smaller amplitude and slower rate of rise of the potential of the more distant electrode E_2. Calibration, 50 μV; 1 msec. [From Loewenstein and Rothkamp (1958).]

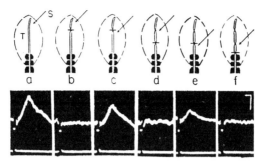

FIG. 4.15   Mechanoelectric conversion after partial compression of the nonmyelinated ending. A subthreshold mechanical stimulus of constant strength is applied to the inner core of a decapsulated corpuscle (a) to (f); only in (a), (c), and (e) is a generator response detected. Lower beam signals relative magnitude and duration of mechanical stimuli; upper beam, the electric activity of the ending led off the myelinated axon (see text). The arrows of the diagram indicate the zone of application of mechanical stimuli. The horizontal lines across the nerve ending T indicate the central boundary of compressed area of ending. At (a), generator response of the intact ending; at (b), after compression of a distal portion of ending; at (c), after moving stimulus application point beyond the compressed zone; at (d), after compressing a zone located centrally with respect to stimulus application point; and at (f), after compression of entire length of ending. Calibration, 1 msec.; 25 μV. [From Loewenstein and Rothkamp (1958).]

Loewenstein and Rothkamp (1958) developed a circuit model, based on the physical observation of the Pacinian corpuscle core. This is shown in Fig. 4.16, where the dashed lines labeled U enclose the suggested circuit of a receptor site. Many of these identical receptor circuits are connected in series through resistances r of the membrane of the core and r_i of the axoplasm, or inner substance of the core. r_s and c_s are the resistance and capacitance between the nonmyelinated ending and the myelinated axon, respectively. Within a given receptor site U, the transmembrane potential is E, the transmembrane capacitance is C_U, and at rest the transmembrane resistance is r_U. When a site is activated, there is a change equivalent to closing an internal switch S and putting in a parallel bypass resistance r_a whose resistance is much lower than r_U. The investigators assumed $r_a = r_U/b$, where b is at least 10. The activation of any site U shunts E with r_a and causes a current to flow through point C of the voltage measuring instrument (upper part of Fig. 4.16). Activation of further sites increases the current flow through C and, hence, the change in potential ΔV of the instrument. If the total number of sensitive sites U is m, and n are activated, define the fraction activated as $x = n/m$. If n sites are activated, the total shunting resistance of the circuit R_T is the parallel sum of the r_a

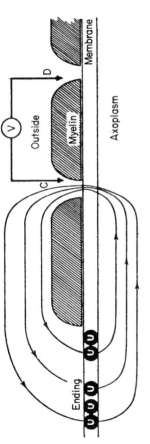
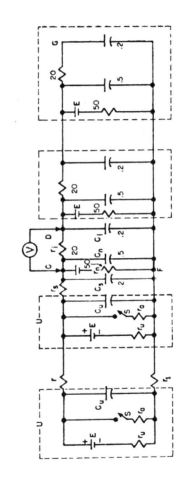

FIG. 4.16 Electric model of a receptor. Resistance values in megohms; capacitance in picofarads. The transmembrane potential E across a receptor site U is assumed to be 100 mV at rest; it is assumed to drop to a fraction of this voltage when the unit is stimulated mechanically. See text for notations and a detailed description. The upper diagram represents some morphological aspects of the receptor. U represents one of many receptor sites assumed to be distributed uniformly over the membrane. These sites do not necessarily imply discontinuity in the membrane structure; they imply only the existence of multiple receptive regions that can be activated independently of each other. Current is represented as flowing between active receptor sites of the ending and the first Ranvier node C. The potential difference between the first (C) and second node (D) is calculated for the equivalent points of the analog. [From Lowenstein (1959).]

whose switches have been closed

$$\frac{1}{R_T} = \sum_{i=1}^{n} \frac{1}{r_{a_i}} = \frac{n}{r_a} \tag{4.1}$$

The sum of all of the parallel capacitances C_T is

$$C_T = \sum_{i=1}^{m} C_{U_i} = mC_U \tag{4.2}$$

and the circuit of Fig. 4.16 with some of the switches closed can be represented by Fig. 4.17.

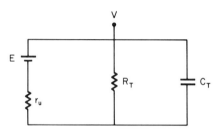

FIG. 4.17   Equivalent circuit of Fig. 4.16.

The total current is given by the equation

$$I = C_T \frac{dV}{dt} + \frac{V}{R_T} + \frac{Vn}{R_T}(V - E) \tag{4.3}$$

and substituting Eqs. (4.1) and (4.2) yields

$$I = mC_U \frac{dV}{dt} + \frac{n}{r_a}V + \frac{m}{r_U}(V - E) \tag{4.4}$$

Substituting $x = n/m$, $r_a = r_U/b$, and letting $I = 0$, since essentially no current flows, we may write

$$\frac{dV}{dt} + \frac{V}{C_U r_U}(bx + 1) = \frac{E}{C_U r_U} \tag{4.5}$$

This equation can be solved by the method of the integrating factor (Vol. I, Appendix Section D3) to yield the solution

$$V = \frac{E}{bx + 1} + K \exp\left(-\frac{bx + 1}{C_U r_U}t\right) \tag{4.6}$$

where K is a constant to be evaluated from the boundary condition. It is seen from the circuit that at $t = 0$ the voltmeter will read the potential E of the battery; thus, at $t = 0$, $V = E$, and substituting this boundary condition into the above solution one obtains

$$K = E - \frac{E}{bx + 1}$$

and therefore

$$V = \frac{E}{bx + 1} + E \exp\left(-\frac{bx + 1}{C_U r_U} t\right) - \frac{E}{bx + 1} \exp\left(-\frac{bx + 1}{C_U r_U} t\right) \quad (4.7)$$

For conciseness of writing, define $R = r_U/(bx + 1)$. Also note that any change in the voltage ΔV that occurs as sites are activated will be the new voltage V minus the initial voltage E or

$$\Delta V = V\text{-}E = \frac{E}{bx + 1} + Ee^{-t/RC_U} - \frac{E}{bx + 1} e^{-t/RC_U} - E$$

$$= \frac{E}{bx + 1} + \frac{(bx + 1)}{(bx + 1)} Ee^{-t/RC_U} - \frac{E}{bx + 1} e^{-t/RC_U} - \frac{(bx + 1)}{(bx + 1)} E$$

$$= \frac{E}{(bx + 1)} \left[1 + bxe^{-t/RC_U} + e^{-t/RC_U} - e^{-t/RC_U} - bx - 1\right]$$

$$\Delta V = -\frac{Ebx}{bx + 1} (1 - e^{-t/RC_U}) \quad (4.8)$$

The minus sign shows that as sites are activated the change in voltage ΔV is negative. It is further seen that if no sites are activated, $x = 0$ and there is no voltage change.

This solution is a saturating exponential, the curve of which has been shown in Vol. I, Fig. 7.10. The time for saturation has been given by Loewenstein as about 1 msec, after which the equilibrium value at $t = \infty$ is approached. It is seen in Eq. (4.8) that at $t = 0$

$$\Delta V = Ebx/(bx + 1) \quad (4.9)$$

This function, with an appropriate constant multiplier $-k$ for circuit potentials encountered in the measurements, is plotted as the solid line of Fig. 4.18a, in which the abscissa is the x value ($x = n/m$, the fraction of activated sites) given in percentage. Remarkably good agreement is seen for this model. What is more significant is the experimental data of Fig. 4.18b. This is a plot of the observed generator potential of an intact Pacinian corpuscle versus stimulus strength by a mechanical oscillator. However, although the electrical analog of the transducer model appears to have been elucidated and tested by experiment, the fundamental biochemical model of the transducer mechanism remains elusive.

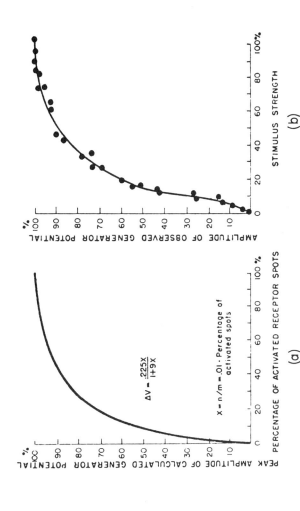

FIG. 4.18 (a) The calculated generator potential as a function of the number of active spots. Peak response $\Delta V = 0.225\ x/(1 + 9x)$, as calculated from Eq. (4.9), is plotted against the percentage of activated receptor spots ($x = n/m = 0.01 \times$ percentage of activated spots). When all available receptor spots are activated (abscissa, 100%) the peak voltage is considered 100%. (b) The observed generator potential as a function of stimulus strength. The amplitude of the generator potential from a typical experiment is plotted against the percentage of stimulus strength. The minimal stimulus strength that produces the maximal generator potential (ordinates, 100%) is considered 100%. [From Lowenstein (1959).]

MECHANICAL TRANSMISSION IN THE PACINIAN CORPUSCLE

In the preceding section, experiments were described which showed that the nonmyelinated nerve ending of the Pacinian corpuscle behaved as a transducer, in that a potential was generated whose magnitude was a function of the area deformed. When there was sufficient deformation, the generated potential reached a critical magnitude which caused an action potential to propagate along the myelinated fiber. It is seen in Fig. 4.15 that upon application of the deformation, there is a rapid rise in potential followed by a decay as the system adjusts to steady pressure. The role of the lamellae, arranged as onion layers, around the core of the Pacinian corpuscle was analyzed by Loewenstein and Skalak (1966), which resulted in rather comprehensive understanding of transient effects in layered biological transducers. The results also compared favorably with the experimental strain measurements on Pacinian corpuscles by Hubbard (1958). The details of the analysis, given in the article by these investigators, is beyond the scope of this book, but the principles and some of the results will be outlined.

The lamellae are considered to be elastic membranes, ovoid in shape, attached together at the ends and with a viscous fluid between the layers. Because of the attachment at the ends, the outer layers are more spheroidal in shape than the inner layers. This is schematically illustrated in Fig. 4.19, in which only a few lamellae are shown for simplicity.

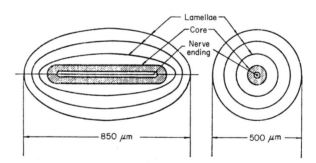

FIG. 4.19 Diagram of a Pacinian corpuscle in longitudinal and transverse sections showing the principal elements of its structure. [From Loewenstein and Skalak (1966).]

To illustrate the physical principle involved consider, as did Loewenstein and Skalak, the behavior of a balloon filled with an incompressible fluid. If the balloon is subjected to an axial compression, it will distort by increasing its diameter along the axis perpendicular to the axis of compression. This distortion increases its volume to compensate for the loss of volume caused by the compression. The elastic wall of the balloon is stretched in this position and acts as a restoring force when the compression is removed. Next, consider that this balloon has a rigid sphere inside, as in Fig. 4.20. If compression of the outer balloon is applied slowly, the fluid will flow easily from the axis of compression to the region of distortion. However, if the compression is applied rapidly, the viscosity of the fluid causes a resistance to rapid flow and some of the compressive force is transmitted to the inner sphere. The amount of transmitted force will be a function of the velocity of compression. Experiments on the Pacinian corpuscle have shown, however, that a static compression on the outer lamella can be transmitted to inner lamellae. Consider the rigid inner sphere of Fig. 4.20 replaced with another balloon filled with the same fluid. While a rapid axial compression on the outer balloon would be transmitted to the inner one because of the fluid viscosity effect, a static compression would not. Therefore, an elastic element connecting the walls of the balloons must be hypothesized. This is most readily visualized as small springs which do not impede the flow of the fluid. Thus, each lamella has three mechanical elements: (1) an elastic one which connects the membrane of the lamella to the ends of the corpuscle, (2) an elastic element which connects each lamella to its adjacent one, and (3) a viscous element which arises from the resistance of the fluid between two lamellae to rapid displacement. This mechanical analog is represented for a cross section of the corpuscle in Fig. 4.21. The treatment of simple viscoelastic assemblies was discussed in Vol. I, Chapter 2. Analogous elements of an electric circuit were given, and Loewenstein and Skalak solved the equation for the equivalent electric circuit of Fig. 4.22 on a computer.

From the model, it is evident that the corpuscle would be sensitive to distortional strain but not to hydrostatic stress, and both calculation and experiment showed that this is indeed the situation. The calculations

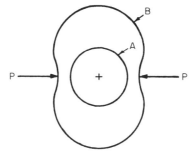

FIG. 4.20 Lamellated model. An inflated balloon A is suspended inside another inflated balloon B. B may be compressed by the forces P without distorting A. [From Loewenstein and Skalak (1966).]

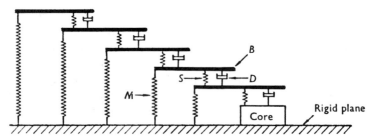

FIG. 4.21   Mechanical analog of the Pacinian corpuscle. Lamella compliance is represented by springs M; radial compliance by springs S; and the fluid resistance by dashpots D.
The lamellae are indicated by the plates B. [From Loewenstein and Skalak (1966).]

showed, however, that when the corpuscle is subjected to a static compression in one direction, less than 3% of the pressure reaches the inner core.
This is because the compliance, i.e., the reciprocal of the spring constant,
of the lamellae interconnections, S, is large compared with that of the
lamellae themselves, M. Thus, most of the distortion is absorbed by the
lamellae instead of being transmitted. The result of such a calculation is
shown in Fig. 4.23. In this calculation, 30 lamellae were used with the
radius of the outer one taken as 255 μm. An initial static compression of 20
μm was applied to the outer lamella and it is seen that the pressure at 20
μm, the core radius, had decreased by about two orders of magnitude.
Thus, although static pressure is detectable by the core, it is small.

During dynamic compression this is not the case, for the fluid viscosity
sets up pressures which are transmitted through the corpuscle. Figure 4.24b
shows both displacement D and velocity V of the displacement versus time
as the outermost lamella was compressed 20 μm at a steady rate for 2 msec
and then held constant. Figure 4.24a shows the pressure versus time on the
outer lamella, the lamella at 0.75 of corpuscle radius (190 μm), and at the
core. It is seen that there is a prompt onset and rapid transmission of the

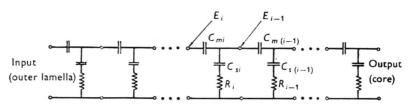

FIG. 4.22   Electrical analog of the capsule of the Pacinian corpuscle. Lamella compliance
is represented by capacitance C_{mi}, the radial spring compliance by capacitance C_{si}, and the
fluid resistance by the electrical resistance R_i. [From Loewenstein and Skalak (1966).]

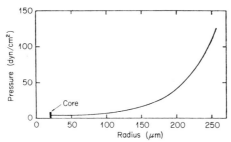

FIG. 4.23 Computed static pressure on lamellae versus radius for a static compression of 20 μm applied to the outermost lamella. [From Loewenstein and Skalak (1966).]

compression while it is being applied. At the end of 2 msec, when the compression is static, the pressure decays to the static values of Fig. 4.23, except for a transient decompression effect at the core.

As expected in this model, a steady compression suddenly released will produce the negative of the above phenomena since the viscous flow must readjust to the new configuration of the cell. This off response was also calculated and an equivalent generator pulse was measured.

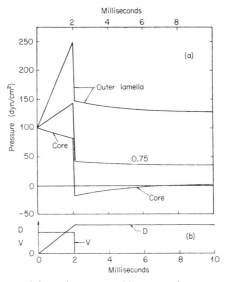

FIG. 4.24 Time course of dynamic pressure: (a) computed pressure on outermost lamella, at 0.75 of corpuscle radius and on the core against time for total compression of 20 μm applied uniformly in 2 msec and then maintained. (b) displacement D and velocity V of outermost lamella. [From Loewenstein and Skalak (1966).]

The experiments and theory briefly described show that the principles of static and velocity responses to applied stresses by layered mechanorecep- tors is now reasonably well understood.

REFERENCES

Adrian, E. D., and Zotterman, Y. (1926). The impulses produced by sensory nerve endings. Part 3, Impulses set up by touch and pressure, *J. Physiol.* **61**, 465.

Andres, K. H., and von Düring, M. (1973). Morphology of cutaneous receptors, *in* "Handbook of Sensory Physiology," Vol. II, "Somatosensory System" (A. Iggo, ed.). Springer-Verlag, Berlin and New York.

Burgess, P. R., and Perl, E. R. (1967). Myelinated afferant fibers responding specifically to noxious stimulation of the skin, *J. Physiol.* **190**, 541.

Burgess, P. R., and Perl, E. R. (1973). Cutaneous mechanoreceptors and nociceptors, *in* "Handbook of Sensory Physiology" Vol. II, "Somatosensory System" (A. Iggo, ed.). Springer-Verlag, Berlin and New York.

Darian-Smith, I., Johnson, K. O., and Dykes, R. (1973). "Cold" fiber population innervating palmar and digital skin of the monkey: responses to cooling pulses, *J. Neurophysiol.* **36**, 325.

Dickhaus, H., Sassen, M., and Zimmermann, M. (1976). Rapidly adapting cutaneous mechanoreceptors (RA): coding, variability and information transmission, *in* "Sensory Functions of the Skin in Primates," Wenner-Gren Center Int. Symp. Ser., Vol. 27. Pergamon, Oxford.

Hensel, H. (1973). Cutaneous thermoreceptors, *in* "Handbook of Sensory Physiology" (A. Iggo, ed.), Vol. II. Springer-Verlag, Berlin and New York.

Herrick, C. J. (1931). "An Introduction to Neurology." Saunders, Philadelphia, Pennsylvania.

Hubbard, J. S. (1958). A Study of rapid mechanical events in a mechanoreceptor, *J. Physiol.* **141**, 198.

Iggo, A. (1969). Cutaneous thermoreceptors in primates and subprimates, *J. Physiol.* **200**, 403.

Iggo, A. (1977). Cutaneous and subcutaneous sense organs, *Brit. Med. Bull.* **33**, 97.

King, J. S., Gallant, P., Myerson, V., and Perl, E. R. (1976). The effects of anti-inflammatory agents on the responses and the sensitization of unmyelinated (C) fiber polymodal receptors, *in* "Sensory Functions of the Skin in Primates" (Y. Zotterman, ed.), Wenner-Gren Center, Int. Symp. Ser., Vol. 27. Pergamon, Oxford.

Kumazawa, T., and Perl, E. R. (1977). Primate cutaneous sensory units with unmyelinated (C) afferent fibers, *J. Neurophysiol.* **40**, 1325.

Loewenstein, W. R. (1959). The generation of electrical activity in a nerve ending, *Ann. N. Y. Acad. Sci.* **81**, 367.

Loewenstein, W. R., and Rothkamp, R. (1958). The sites for mechano-electric conversion in a Pacinian corpuscle, *J. Gen. Physiol.* **41**, 1245.

Loewenstein, W. R., and Skalak, R. (1966). Mechanical transmission in a Pacinian corpuscle, an analysis and a theory, *J. Physiol.* **182**, 346.

Lynn, B. (1975). Somatosensory receptors and their CNS connections, *Ann. Rev. Physiol.* **37**, 105.

Lynn, B. (1977). Cutaneous hyperalgesia, *Brit. Med. Bull.* **33**, 103.

Miller, M. R., Ralston, H. J., and Kasahara, M. (1960). The pattern of cutaneous innervation of the human hand, foot and breast, *in* "Advances in Biology of Skin" (W. Montagna, ed.), Vol. 1, "Cutaneous Innervation." Pergamon, Oxford.

Montagna, W., and Parakkal, P. F. (1974). "The Structure and Function of the Skin," 3rd ed. Academic Press, New York.

Penfield, W., and Rasmussen, T. (1950). "The Cerebral Cortex of Man." Macmillan, New York.

Perl, E. R., Kumazawa, T., Lynn, B., and Kenins, P. (1976). Sensitization of high threshold receptors with unmyelinated (C) efferent fibers, *Prog. Brain Res.* **43**, 263.

Sinclair, D. (1967). "Cutaneous Sensation." Oxford Univ. Press, London and New York.

Vallbo, A. B., and Hagbarth, K. E. (1968). Activity from skin mechanoreceptors recorded percutaneously in awake human subjects, *Exp. Neurol.* **21**, 270.

Vallbo, A. B., and Johansson, R. (1976). Skin mechanoreceptors in the human hand: neural and psychophysical thresholds, *in* "Sensory Functions of the Skin in Primates" (Y. Zotterman, ed.), Wenner-Gren Symp. Ser., Vol. 27. Pergamon, Oxford.

Van Hees, J., and Gybels, J. (1972). Pain related to single afferent C fibers from human skin, *Brain Res.* **48**, 397.

CHAPTER 5

Audition

INTRODUCTION

The ear is both a remarkably sensitive and rugged instrument. It responds to sounds whose frequency may vary over ten octaves, while the eye responds to about one octave. It responds to air vibrations whose amplitude is the order of atomic size and to sounds ten million million times greater. Its response, however, is not proportional to either frequency or intensity and, although the response has been mapped, the physiological mechanisms of this behavior have not yet been determined.

The word response can be deceptive in its usage. The ear is not a passive transducer which simply converts mechanical to electrical energy. There are, in fact, three amplification stages. The first is the hydraulic piston action of the ossicles, the small bones of the middle ear. The second is a shearing mechanism which occurs between two membranes of the inner ear. The third is an electric amplifier in which a resistance proportional to displacement varies the current flow to the auditory nerves. These will be discussed in turn.

It is now known that the frequency, or pitch, sensitivity of the ear is not that proposed by Helmholtz in 1885 (1954), in which each part of the inner ear is sensitive to a given frequency. Instead, although there is some mechanical resonance within the inner ear, these resonances are broad in frequency. While some neurons in the auditory cortex are frequency selective, most are not. It now appears that a complex of neurons, including inhibitory ones, determine the sharp frequency sensitivity. The origin of such frequency sensitivity is not yet solved, although there have been many noble attempts. Reviews of these theories with references are given in the books by von Békésy (1960), Littler (1965), and Wever and Lawrence (1954). For a more recent proposal see Zwislocki and Kletsky (1979). It is not the purpose of the present chapter to review the mathematically sophisticated theories which are either incomplete or incorrect. Instead, there will be a review of the methods of physics, with some physiology, which have been used to determine many of the facts which we know about the hearing mechanism. This will be followed by a brief discussion of deafness and corrective surgery which has been based on these facts. For a more mathematical treatment of many of the topics in this chapter see Dallos (1973).

UNITS OF MEASUREMENT

If a sound wave is sinusoidal, there are two measurements which can, in principle, be made: the intensity and the root mean square (see the Appendix) pressure. The root mean square (rms) pressure is defined as

$$p_{rms} = p_{max}/\sqrt{2} \tag{5.1}$$

where p_{max} is the maximum of peak pressure of the sound wave. The connection between intensity, usually given in watts per square meter (although in older papers as watts per square centimeter), is (see Vol. I, Eq. (8.15′))

$$I = p_{max}^2/2\rho c = p_{rms}^2/\rho c \tag{5.2}$$

where ρ is the air density and c the velocity of sound in air.

Since the ear is sensitive to many orders of magnitude of sound intensity, the decibel scale is used. The level difference in bels is the logarithm to base 10 of the ratio of intensities

$$N(\text{bels}) = \log_{10}(I/I_0) \tag{5.3}$$

and to expand the scale decibels (dB) are used where

$$N(\text{decibels}) = 10\log_{10}(I/I_0) \tag{5.4}$$

Note that 1 bel = 10 dB and 2 bel = 20 dB, etc. By substitution of Eq. (5.2), we may equivalently write

$$N(\text{dB}) = 10 \log_{10}\left(p_{\text{rms}}^2/p_{0\,\text{rms}}^2\right)$$

or

$$N(\text{dB}) = 20 \log_{10}\left(p_{\text{rms}}/p_{0\,\text{rms}}\right) \tag{5.5}$$

It is seen that for Eq. (5.5) to be finite, $p_{0\,\text{rms}}$ must have a finite value for the zero pressure level. This is chosen at the human audio threshold. As will be shown, this threshold varies with frequency and so the threshold was selected at 1000 Hz. The threshold level was chosen in order to avoid negative numbers in relative sound intensities. The selected threshold is 10^{-12} W/m², which at standard temperature and pressure corresponds through Eq. (5.2) to approximately 0.0002 μbar $= 2 \times 10^{-5}$Pa $= 2 \times 10^{-5}$ N/m².

SENSITIVITY OF THE EAR

When a number of persons with normal hearing are tested to learn their sensation of equal loudness upon comparing sounds of different single frequencies, the average of the results can be plotted as in Fig. 5.1. The bottom curve represents threshold, or zero loudness. Note here that the psychophysiological quantities *loudness* and *pitch* correspond to the physical quantities *intensity* and *frequency*, respectively. It is seen in Fig. 5.1 that the loudness of a sound depends on both intensity and frequency. While the pitch depends chiefly on frequency, experiments have shown that it is somewhat dependent on intensity. The upper curve represents the threshold of pain, and sound intensities above this can be physiologically damaging to the ear. About 5 dB below this threshold there is a tickling sensation in the ear. The sound level in some metal stamping factories is often above 130 dB, which results in damage to the ears of workers (Kryter 1970). Sound levels in some discotheques and even in the New York City subway have been measured to be at this level.

Figure 5.1 shows that the normal ear can hear sounds of frequencies in the range 20–20,000 Hz with an intensity range of 120 dB (12 orders of magnitude of intensity). The wavy appearance of some of the loudness contours would not be present if the measurements were made at the eardrum. The fluctuations occur because of cavity resonance of the ear canal, increasing the sensitivity around 4000 Hz and decreasing it around 8000 Hz.

The contours of equal loudness of Fig. 5.1 are not parallel to each other. Therefore, if there is a complex sound consisting of several frequencies, the *quality*, or relative intensity of the frequencies, will change when the overall

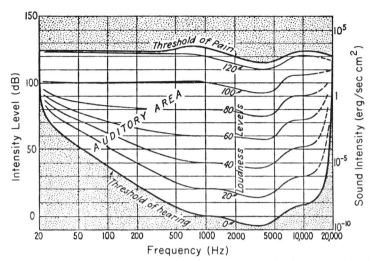

FIG. 5.1   Contours of equal loudness plotted against intensity level and frequency for the average human ear. Loudness levels are in dB above threshold at 1000 Hz. Unshaded area is the auditory range over which sound can be heard. [From P. M. Morse (1948).]

intensity is changed. Suppose a recording is made at a sound intensity of 100 dB of some complex sounds centered about the frequencies 100 and 1000 Hz. If it is then played back at an intensity of about 20 dB, the sounds centered about 100 Hz would not be heard at all. Thus, a recording when played no longer has "high fidelity" and the tone control on the record player must be adjusted with the loudness control to the individual listener's taste. In other words, there is no such thing as a high fidelity recording and record player unless the record is played at the sound intensity at which it was recorded.

Let us now consider the amplitude of motion of air molecules at 20°C and atmospheric pressure when vibrating with a sound intensity of 10^{-12} W/m². In the preceding section, we saw that this corresponds to an rms pressure of 2×10^{-5} N/m² or a $p_{max} = 2.8 \times 10^{-5}$ N/m². We may calculate the amplitude of the molecular vibration at any frequency from the relation (Vol. I, Eq. (8.14))

$$p_{max} = \omega \rho c \xi_0 \qquad (5.6)$$

where $\omega = 2\pi$ frequency, ρ is the density, c the velocity of sound, and ξ_0 the amplitude of vibration. The values of c and ρ at 20°C are $c = 343$ m/sec and $\rho = 1.21$ kg/m³. Substituting these values into Eq. (5.6), we obtain for 1000 Hz $\xi_0 = 1.07 \times 10^{-11}$ m $\cong 1 \times 10^{-9}$ cm. Note that the diameter of a hydrogen atom is 10^{-8} cm, so that the ear can detect motion of one-tenth that diameter.

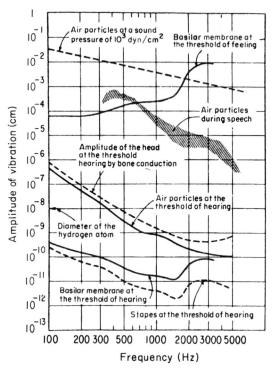

FIG. 5.2   The vibration amplitude of different parts of the ear for the threshold of hearing and of air particles at different sound levels compared with the diameter of a hydrogen atom. [From von Békésy (1962).]

Figure 5.2 shows the amplitudes of vibration versus frequency of parts of the ear and of air molecules for comparison. The sensitivity of the ear is at the noise level of Brownian motion of molecules, and this problem has been addressed by Naftalin (1965) and de Vries (1948). Since Brownian motion is the physical lower limit of sensitivity, it appears that evolution has brought the sense of hearing to its limit and it will not be improved upon in the future. In the next chapter, it will be shown that the eye also is at its evolutionary limit of development.

ANATOMY OF THE MIDDLE EAR

The middle ear is a rather complex three-dimensional structure which is difficult to visualize from two-dimensional illustrations. However, since we are interested in the mechanics of its operation more than its morphology, schematic drawings will be used when needed.

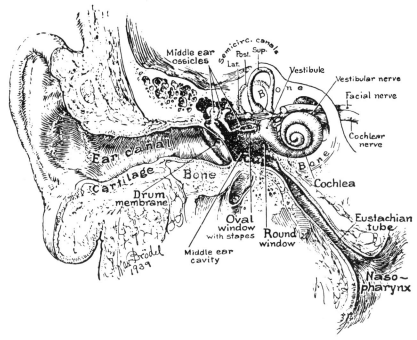

FIG. 5.3. A drawing of the anatomy of the human ear. [Drawing by Max Brödel. From Brödel (1946).]

A cross-sectional drawing of the ear is shown in Fig. 5.3. The ear canal terminates at the drum membrane. Attached to the middle ear side of the drum is the first of a series of three small bones, or *ossicles*, which are linked together.

The middle ear contains air which is maintained at external pressure by the *eustachian tube*. This tube is normally closed to prevent foreign matter from entering the inner ear but opens upon swallowing to adjust the pressure. This opening and closing is the "click" sound that we can hear when swallowing. When there is considerable pressure difference between the middle ear and the outside, the air pressure differential forces air through the eustachian tube. For this reason, we experience a small sound when going up in altitude; air from the middle ear is forcing its way out through the eustachian tube.

To the right of the middle ear in Fig. 5.3 is the *cochlea* which, as its name suggests, is a snail shell type of structure. This is actually the tissue part of a spiral cavity in bone so that the cochlea does not bend or yield under pressure. The semicircular canals illustrated here are part of the *vestibular* (balance) sense and are not involved in the hearing process. The cochlea contains two types of fluid lymphs and has two windows in its bony structure, both facing the middle ear and both covered with a membrane. The *oval window* is the upper one and the third bone (the *stapes*, or stirrup) of the three ossicles is attached to its membrane. The *round window* is the lower one and its membrane simply faces the middle ear. The membranes over these two windows are flexible. Since the cochlea is not flexible and the fluid is incompressible, a pressure wave transmitted from the drum through the stapes to the oval window cannot compress the fluid, so the round window will bulge out slightly. A more detailed view of the oval and round windows is shown in Fig. 5.4, in which the cochlea lies to the left. The arrows of the compression wave start at the oval window (stapes removed) and end at the round window.

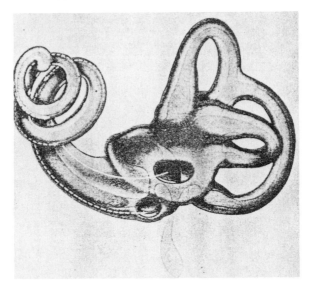

FIG. 5.4.   The membranous labyrinth (within the bony labyrinth) (left ear); the cochlea is on the left and the three semicircular canals of the vestibular (balance) system on the right. The arrows indicate the direction in which sound disturbance starts at the oval window, or vestibule. The solid line traces the scala vestibuli to the apex (helicotrema) and the dashed line to the scala tympani back to the round window. [From "Gray's Anatomy: The Unabridged Running Press Edition of the American Classic," by Henry Gray. Copyright © 1978 Running Press. Reprinted courtesy of Running Press, 38 South 19th Street, Philadelphia, Pa. 19103.]

The three ossicles are shown in Fig. 5.5 in their arrangement in the right ear viewed from the front, although they have been rotated slightly to the right. On the left is the *malleus*, or hammer, in the center in the *incus*, or anvil, and on the right is the *stapes*, or stirrup. The malleus is attached to

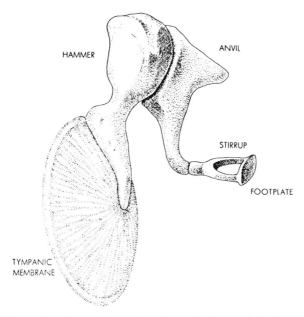

FIG. 5.5. The eardrum (tympanic membrane) and the three ossicles; hammer (malleus), anvil (incus) and the stirrup (stapes) in the middle ear. The footplate of the stirrup is attached to the oval window. [From G. von Békésy, The ear, *Sci. Am.* **197** (Aug.), 232 (1957). Copyright © 1957 by Scientific American, Inc. All rights reserved.]

the inner part of the eardrum, the *tympanic membrane*, as shown in Fig. 5.5. Omitted from this drawing is a small length of bone attached to the malleus called the *long process*, Fig. 5.6. This long process is in children only, and it appears to be absorbed in adults. It has been suggested that this long process helps achieve balance in the rotational motion of the malleus, and the amount not required is absorbed. The three ossicles are attached to each other by five ligaments, and the stirrup is attached to the oval window.

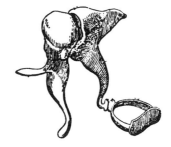

FIG. 5.6 Bones of the middle ear. The long process of the malleus is the horizontal bone on the left. [From Helmholtz (1954).]

The ossicles are suspended at several points such that they act as a complex lever unit. As the eardrum moves to the right in Fig. 5.5, the stirrup, and hence the oval window, also moves to the right. The attachment of the eardrum to the malleus keeps the membrane of the drum in constant tension, so that a disturbance at any point on the drum causes motion. This arrangement of the ossicles between the membranes of the eardrum and the oval window amplifies the pressure of a disturbance on the drum with the mechanism of a hydraulic press. The amplitude of the handle of the malleus, i.e., that part at the center of the drum, is about 1.3 times that of the stapes. The surface area of the drum is about 55 mm^2, while that of the oval window is about 3.2 mm^2, giving an area ratio of 17. Thus, the ratio of the transmitted pressure from the eardrum to the oval window is 1.3 $\times$ 17 = 22. Although not yet described here, the cochlea is filled with fluid which must be set in motion by the membrane of the oval window. Since fluids have much greater inertia than air, increased force must be applied, otherwise much of the sound energy would be reflected from the eardrum. The hydraulic press arrangement acts as an impedance matching device.

Any oscillatory system, such as the motion of the ossicles, has a resonant frequency, that is, a frequency at which it oscillates with a minimum of input energy. In the human ear, this resonant frequency range is fairly broad and spans the range 700–1400 Hz. This is the frequency range of maximum auditory acuity of the ear. However, the ligaments which attach the ossicles are viscoelastic and thereby suppress severe oscillations.

There are two small muscles in the middle ear, both attached to the inner wall of the cavity, or *tympanum*, of the middle ear. One, the *tensor tympani*, attaches to the handle of the malleus. The other, the *stapedius*, is attached to the neck of the stapes. The tensor tympani draws the membrane of the tympanum inward and thus heightens the tension for the motion of the malleus. The stapedius draws the head of the stapes backward and thus causes the base of the bone to rotate on a vertical axis through its own center. In doing this, the back part of the base would press inward on the oval window membrane, while the fore part would be drawn from it. This motion probably compresses the fluid in the vestibule.

It has been proposed that these muscles act reflexively like the blinking of the eyelids. The increased tension of the system decreases the mechanical impedance matching for the transmission of low frequencies. It is estimated that the low frequency energy transmission reduction can be as high as 40 dB. The reflex action has been demonstrated to occur for clicks and loud noises. It also occurs during swallowing and yawning. A yawning student will note the diminished sound of a teacher's voice. The protective action of this acoustic reflex is limited. It begins in the stapedius about 15 msec after sound stimulation and a bit later in the tensor tympani, although the latency is a function of sound intensity (Towe, 1965). Because of this latency, no protection is available for a sudden sound such as an explosion.

THE COCHLEA

The oval window, discussed in the preceding section, is the entrance of acoustical disturbances to the *inner ear*. The inner ear is enclosed in the *labyrinth*. The labyrinth is actually two parts: one the *osseus labyrinth*, a series of cavities within the temporal bone, the other the *membranous labyrinth*, a group of sacs and ducts within the bony cavities. An illustration of the membranous labyrinth is shown in Fig. 5.4, where the cochlea lies to the left. The loops on the right are the *vestibular* or *semicircular canals*, by which we maintain our sense of balance. Indicated in this figure is the path of an acoustical disturbance, which originates at the oval window and finally causes a displacement of the round window, the lower opening of Fig. 5.4. The length of the cochlea is about 3.5 cm and has 2.5 to 2.7 complete revolutions in its coils.

A cross-sectional view of the cochlea is shown in Fig. 5.7, and a more schematic view is shown in Fig. 5.8. There is a bony shelf extending partially into the spiral labyrinth. From this shelf to the outer wall stretches a tough membrane called the *basilar membrane*, which is attached on one side to a bony shelf and to the other by the *spiral ligament*. It continues throughout the spiral, separating the canal into two passages up to the apex, called the *helicotrema*, where the passages connect. The lower passage is opposite the round window and is called the *scala tympani*. Extending diagonally above the basilar membrane is a much more delicate one called *Reissner's membrane*. This extends along the basilar membrane to the helicotrema but is not open at either end. It forms the cochlear duct, which has the name *scala media*. The upper duct, called the *scala vestibuli*, has its lower end opposite the oval window and is the passage for the acoustic impulses relayed from the middle ear. Attached to the basilar membrane within the scala media is the *organ of Corti*, in which acoustical displacements are converted to electrical signals. This will be discussed in detail later. The scala media is filled with a fluid called *endolymph*. The scala tympani and scala vestibuli are filled with a watery fluid called *perilymph*. The tension in the basilar membrane is equal in all directions in its plane. This was shown by von Békésy (1960) who photographed depressions made by a point stylus in the membrane of cadavers. The depression is circular, whereas it would be elliptical for the case of unequal

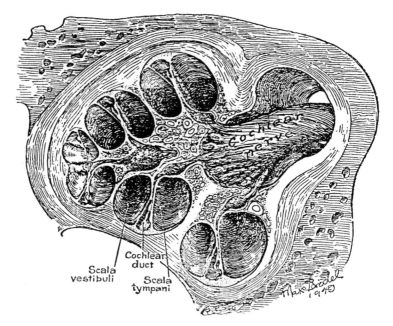

FIG. 5.7 Cross-sectional drawing of the cochlea. The cochlear duct contains the scala media and organ of Corti. [From Brödel (1940). Original illustration in Collection of Art as Applied to Medicine, The Johns Hopkins School of Medicine.]

tension. The width of the basilar membrane increases from the stapes to the apex. Although at any point on the basilar membrane the elastic properties are similar, the elastic modulus decreases by a factor of 100 from the stapes to the helicotrema. Elasticity used in this sense is a measure of the volume elasticity per unit length $\Delta v/\Delta l$; for a fixed pressure from one side of the membrane, its volume displacement per unit length is measured. Because of its dimensions and elastic properties, the

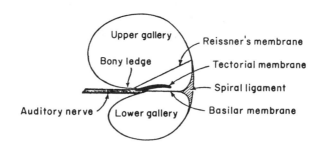

FIG. 5.8 Cross section of the cochlea indicating the membranes important to audition. [From Kinsler and Frey (1962).]

calculation of the vibrational modes of the membrane is not simple. The determination of the modes was done experimentally.

The coil structure of the cochlea is a space-saving device and has little effect on the acoustical properties of the inner ear (von Békésy, 1960). This is known from the acoustical response of other animals studied whose ears have all the properties of human ears except that the cochlea is straight. For ease of visualization, we can consider a straight form of the cochlea, Fig. 5.9.

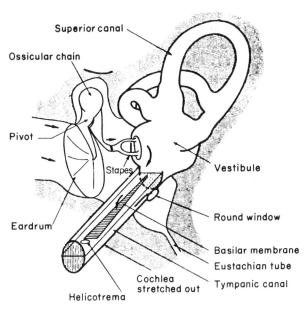

FIG. 5.9   Schematic drawing of the middle and inner ear. The cochlea is shown uncoiled into a straight tube. Arrows show the displacement of the fluid for an inner movement of the eardrum. [From G. von Békésy (1962).]

THE BASILAR MEMBRANE

The pressure waves on the eardrum are transmitted through the ossicles which cause the stapes to move the membrane of the oval window back and forth. This motion at the oval window causes pressure waves in the perilymph of the scala vestibuli, i.e., the passage above the basilar membrane. These acoustical pressure waves set up a pattern of vibrations in the basilar membrane. Such vibrations would not be possible if the perilymph below the membrane were in a completely rigid cavity. However, the membrane covering the round window serves as an outlet for the pressure waves and permits the up-and-down motion of the basilar membrane. Such motion is shown schematically in Fig. 5.10.

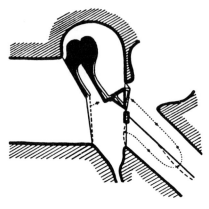

FIG. 5.10 Schematic diagram of portions of the middle and inner ear. Solid lines represent rest positions of the membranes of the eardrum and the oval and round windows. Dashed lines represent their displacement for motion of the eardrum inward. Dotted lines and arrows indicate path of sound waves. [From Stevens and Davis (1938).]

The basilar membrane is most narrow near the stapes and broadens by a factor of three to four as it approaches the apex. As expected, there is a greater amplitude of vibration for low frequencies near the apex and for high frequencies near the stapes (Davis 1953). von Békésy (1943, 1960) showed that, although this is the situation, the vibrations are quite complex and do not exist as standing waves in the basilar membrane. Working with

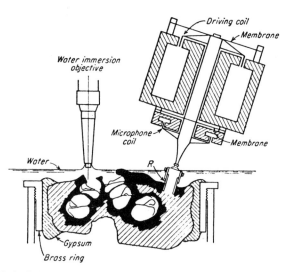

FIG. 5.11 Method of measurement of the amplitude of vibration of the basilar membrane for known displacements of an artificial stapes. [From von Békésy (1949b).]

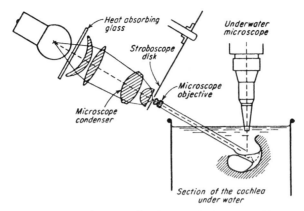

FIG. 5.12   Equipment for stroboscopic illumination of a human cochlear partition. [From von Békésy (1949).]

cadavers, he opened the cochlea at various positions and observed the vibrations of the basilar membrane with a stroboscopic light when the stapes were vibrated at different frequencies, Figs. 5.11 and 5.12. Example data are shown in Fig. 5.13. It is seen that the response is rather wide with a definite maximum that is increasingly closer to the stapes with increasing frequency. When a sudden sound is applied to the stapes, a wave travels along the basilar membrane toward the apex. von Békésy measured the time of arrival of the wave at different positions along the membrane. These data are shown in Fig. 5.14. With the same type of experiment, von Békésy was able to determine at any point the variation of the amplitude

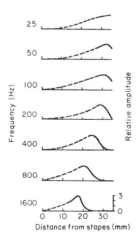

FIG. 5.13   Displacement amplitudes along the basilar membrane for different frequencies. [From von Békésy [1949b).]

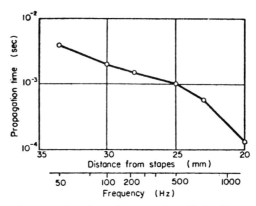

FIG. 5.14 Time of propagation of a pulse wave along the basilar membrane. The upper
scale of the abscissa is the distance from the oval window and the lower scale are the
resonance frequencies of the spatial positions. [From von Békésy (1949b).]

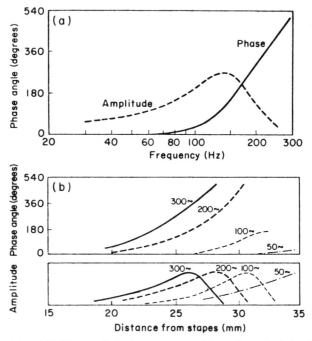

FIG. 5.15 (a) and (b) Phase angle in degrees measured from the oval window for different
frequencies. Dashed curve in (a) and lower curves are corresponding amplitudes. [From von
Békésy (1943).]

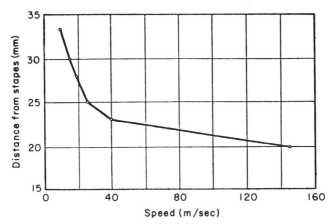

FIG. 5.16   Speed of conduction of a sound wave in the basilar membrane. [From von Békésy (1943).]

of vibration and phase angle difference relative to the stapes. A typical set of data is shown in Fig. 5.15. In resonance vibration of a simple physical system, the phase angle changes from $-\pi/2$ to $+\pi/2$ as the frequency is varied from well below the resonance frequency to one well above it, while at the resonance frequency the phase angle is zero. However, although the lower part of Fig. 5.15 shows that the amplitude for different frequencies has the shape of a resonance curve, the phase angle data show that the system does not undergo a simple resonance phenomena. Such a continuously increasing phase angle from the stapes to apex is characteristic of a traveling wave. We have seen in Fig. 5.14, and it is shown more explicitly in Fig. 5.16, that the velocity of the wave decreases toward the apex, which Ranke (1942) predicted from theoretical calculations based on variations of mass of the basilar membrane. Because of either a variation of mass or elasticity, the traveling wave is damped and, in his experiments, von Békésy was able to observe only two or three crests of the wave from the stapes. A schematic of such observations is shown in Fig. 5.17. For further discussion of the traveling wave model and more complete references, see Wever and Lawrence (1954), Littler (1965), or von Békésy (1960).

There has been some confusion in the literature as to the meaning of the descriptive term "traveling wave" used by Békésy. Traveling waves are well-known physical phenomena, such as surface waves in water or shear waves in a rope. These waves transfer energy from one element of the medium to the next. The experiments of von Békésy showed that the amplitudes of motion of various segments of the basilar membrane were out of phase and that the phase lag increased with distance from the

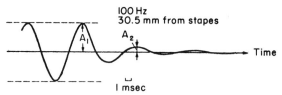

FIG. 5.17   The decay of amplitude in a portion of the basilar membrane 30.5 mm from the stapes, which is being driven at 100 Hz. [From von Békésy (1943).]

stapes. These observations were made with a strobe light which illuminated the basilar membrane through windows drilled through the bony labyrinth. The observed phase lag can be described in terms of a traveling wave. However, in the cochlea, the force is transmitted through the perilymph and is applied directly to each vibrating element of the basilar membrane. This is a different mechanism from the water or rope example. What is happening is that each element of the basilar membrane is executing sinusoidal vibrations, but different elements are executing these vibrations in different phases. This action can be referred to as that of a traveling wave provided that it is simply descriptive without implying anything about the underlying causes (Wever and Lawrence, 1954; von Békésy, 1960).

THE ORGAN OF CORTI

Within the scala media lies the organ of Corti. This extremely sensitive organ serves as both an amplifier and a transducer for converting mechanical into electrical energy. It rests on the basilar membrane, Fig. 5.18, and extends for the full length of it. A fibrous, triangular structure called the *rods of Corti* rests on each basilar membrane fiber. This triangle is rigid and its apex is continuous with an upper rigid membrane called the *reticular membrane*. The sensory parts of the organ of Corti are the *hair cells*. The rods of Corti divide the organ of Corti into inner and outer portions, and the hair cells have the corresponding names, i.e., *inner* (internal) and *outer* (external) hair cells. There is a single row of inner hair cells, numbering about 3500 and measuring about 12 μm in diameter, and three to four rows of external hair cells, numbering about 20,000 and having a diameter of about 8 μm. The hair cells are imbedded firmly in the reticular lamina, while the hairs themselves are imbedded in the surface of the *tectorial membrane*. Therefore, movement of the basilar membrane upward results in a shearing force on the hairs, which causes them to bend one way, while a downward movement causes them to bend in the opposite direction. Thus, a small motion of the basilar membrane gives rise to a large shearing displacement and thereby a further amplification of the sound wave.

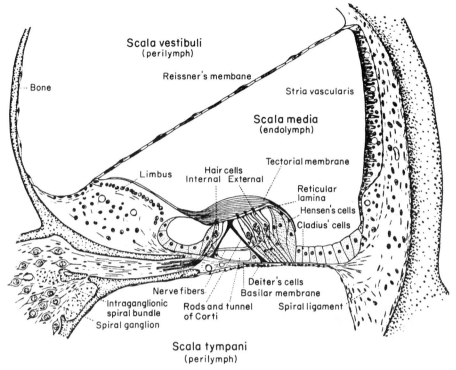

FIG. 5.18   Schematic diagram of the cochlear duct at the second turn of the cochlea of a guinea pig. [From Davis *et al.* (1953).]

HAIR CELLS

Mechanoreceptor cells which have protruding small hairs, or *cilia*, which respond by electrical signals when acted upon by shearing forces, occur frequently in the animal kingdom. Not only are they found in the organ of Corti but also in the vestibular canals and in the organs of fish, which serve to sense low frequency vibrations in the water. Thus, there is a variety of species which can serve as specimens for the detailed examination of these cells.

The guinea pig has an auditory system similar to that of man and has been used extensively for studies of the ear, while fish have been used for some detailed studies of hair cells. An electron micrograph of a section of the organ of Corti of a guinea pig, Fig. 5.19, shows three rows of external hair cells with the hairs fixed in the reticular lamina. The tectorial membrane in which the hairs are imbedded has been removed and with it part

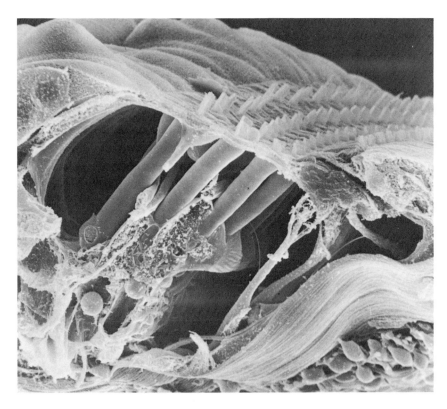

FIG. 5.19 Electron micrograph of a cross section of the organ of Corti showing the external hair cells (magnification 1500 ×). [From Bredberg (1977).]

of the hairs. Thus, only stubs of the hairs are seen in Fig. 5.19, the upper parts having been broken off. A schematic drawing of the arrangement (viewed in the opposite direction) is shown in Fig. 5.20 and more detailed drawings of individual hair cells in Fig. 5.21. Each hair cell synapses to an afferent and efferent nerve fiber which send signals to the central nervous

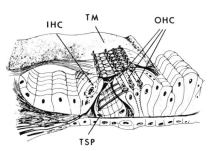

FIG. 5.20 Schematic diagram of the organ of Corti. OHC, outer hair cells; IHC, inner hair cells; TM, tectorial membrane; TSP, tunnel spiral fibers. [From Wersäll et al. (1965).]

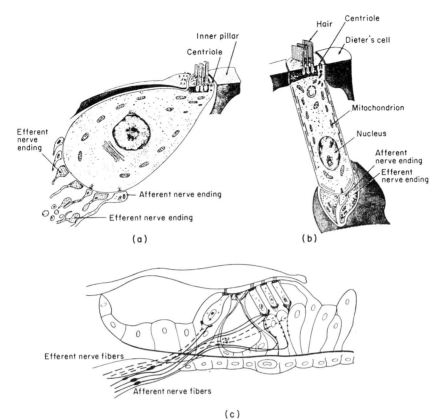

FIG. 5.21   Schematic drawings of (a) inner hair cell and (a) outer hair cell. (c) Location of hair cells in organ of Corti. [From Wersall et al. (1965).]

system and receive feedback. The efferent nerve exists to modify the signal in an as yet unknown way.

A careful study by Flock (1965) of a lateral line receptor in the acoustico-lateralis system of the fish *Lota vulgaris* reveals specific information of the potential changes with motion of the hairs in the hair cells. Electric potential changes were measured in the region surrounding the cell membrane. Figure 5.22 shows a schematic drawing of the excitation potential. The amplitude of the potential is a nonlinear function of the displacement of the hairs. The upper double loop of the schematic is a polar plot of the magnitude of the potential for a given displacement in corresponding directions. It is seen that (1) the magnitude is greatest in one direction, (2) one direction of motion depolarizes the potential and is therefore excitatory, while the other hyperpolarizes and is therefore inhibitory, and (3) the excitatory potential is greater than that of the inhibitory potential for an equal amplitude of displacement. A summary of the effects

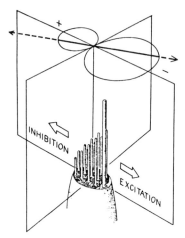

FIG. 5.22   The amplitude of the receptor potential with bending of hair cells in the lateral line canal of a fish. The receptor potential generated by equal amplitudes of displacement are proportional to the distances cut out by the circles. [From Flock (1965).]

of hair motion is illustrated in Fig. 5.23. The upper part of this figure shows the hyperpolarization and depolarization potentials as the hairs are moved. Note that, in agreement with Fig. 5.22, the depolarization potential is the larger for a given amplitude of hair displacement. The signal passing through the afferent nerve has a resting rate of spikes, or action potentials. The depolarization motion increases the rate of the spikes, while hyperpo-

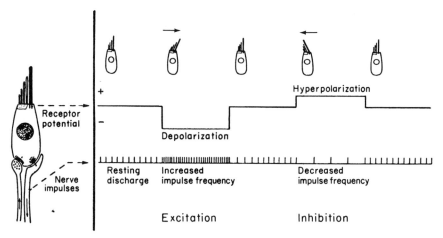

FIG. 5.23   Diagram of proposed model of action potentials versus hair cell displacement. [From Flock (1965).]

larization decreases the rate. It should be emphasized that this schematic of the spike rate is inferred. No experiment has yet been performed which inserts an electrode into the afferent nerve of a single hair cell while it is being moved in controlled ways. What has been accomplished is the measurement of the potential change within and external to the hair cell itself. Thus, while Fig. 5.22 is an accurate portrayal of experimental results, Fig. 5.23 illustrates the probable behavior from our knowledge of other receptor cells for which spike rates have been measured when the potential in a receptor cell reaches a critical value.

ELECTRICAL PHENOMENA OF THE COCHLEA

There are four types of electrical phenomena in the cochlea (Davis, 1957, 1961). These have been studied independently of each other and have different origins. In this section they will simply be defined, while the subsequent four sections will describe detailed experiments on each type which trace the signal from the mechanical transduction to the brain.

Endolymphatic Potential (EP). This is a resting potential difference, about 80 mV positive, of the scala media with respect to the scala tympani, scala vestibuli, or tissues of the head.

Cochlear Microphonic (CM). This is an electrical potential, measured in any of the scalae, which occurs when the basilar membrane is displaced. It can be positive or negative depending upon the direction of membrane displacement, and is proportional to the magnitude of the displacement up to about 5 mV and follows the frequency of oscillation up to about 8000 Hz.

Summating Potential (SP). The summating potential is a change in the endolymphatic potential (EP) in response to acoustic or mechanical stimulation of the inner ear. It is proportional to the intensity of the stimulation, but is unidirectional in sign in contrast to the bipolar response of the cochlear microphonic. Its magnitude is proportional to the root mean square (rms) acoustic pressure, which is a time integral, rather than to the instantaneous displacement. The summating potential behaves as a baseline on which cochlear microphonics are superimposed.

Action Potential (AP). This is the usual series of all-or-none spike potentials observed in nerve transmission and is detected by fine electrodes in the auditory nerves.

In addition to these potentials peculiar to the ear most, if not all, cells have a potential of -20 to -80 mV, negative with respect to the surrounding fluid. This is a general characteristic of body cells.

ENDOLYMPHATIC POTENTIAL

The measurement of the resting potentials in the inner ear was first done accurately by von Békésy (1952) on a guinea pig. A small hole was drilled in the bony wall just above Reissner's membrane, and a micropipette of about 0.3-μm diameter filled with an appropriate electrolyte was carefully inserted. The path of the electrode was followed with a microscope so that known landmarks could be established and related to voltage changes. The second, or indifferent, electrode was similar in construction and was inserted into the perilymph of the scala vestibuli. Thus, the perilymph potential is taken as zero, and the voltages of the probing electrode are relative to that of the perilymph.

The voltages of two separate electrode probings are shown in Fig. 5.24. It is seen that the cells within Reissner's membrane are about -20 mV, as expected of cell interiors relative to external fluids. When Reissner's membrane is penetrated and the electrode is immersed in the endolymph of the scala media, the potential is about $+50$ mV. It is constant throughout the endolymph, as seen in the a and b penetrations of Fig. 5.24. During penetration of Hensen's cells of the cochlea and the cells of Claudius in the basilar membrane, the potential is -40 mV. Penetration of the basilar membrane exposes the electrode to the perilymph of the scala tympani which has a potential of -2 mV.

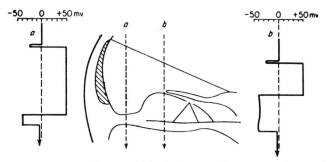

FIG. 5.24   Changes in the dc potential in the downward movement along paths a and b. [From Békésy (1952).]

The potential of the perilymph is not the same along the length of the two scalae. von Békésy (1951a) showed this by drilling a series of holes in the bony wall of the cochlea of a guinea pig. The hole near the footplate of the stapes was taken as zero potential, and Fig. 5.25 shows the potential in the perilymph of the scala vestibuli and scala tympani relative to the zero position. The spiral of the cochlea has been straightened for ease of visualization. These differences have two possible explanations: (1) the salt concentration varies along the scalae, which would give rise to potential differences between the electrodes or (2) there is no salt concentration difference, but an actual electrical current flow. As Békésy has pointed out, the two scalae connect at the helicotrema, and it has been shown that motion of the basilar membrane causes a streaming of the perilymph between the scalae. Thus, mixing clearly occurs and the potential difference must arise from current flow rather than salt inhomogeneities. The implications of this conclusion will be considered later.

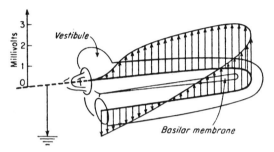

FIG. 5.25 The dc potentials in the perilymph of the vestibular and tympanic scalae. [From von Békésy (1951a).]

COCHLEAR MICROPHONICS

When any part of the inner ear is displaced, the resting potentials change. Extremely delicate experiments have been performed by a number of investigators to ascertain the characteristics of these changes and to locate their origin.

When electrodes are inserted into the scalae vestibuli and tympani and the basilar membrane is pressed by a needle in the scala tympani, the voltage in the scala tympani increases relative to that of the scala vestibuli. This increase is approximately proportional to the pressure of the needle against the basilar membrane (von Békésy, 1960).

It was proposed that the potential changes occurred in the endolymph and that Reissner's membrane, being so thin as to be transparent, would enable potential changes to be capacitively imaged in the perilymph. von Békésy performed an ingenious experiment in measuring the potential change in the perilymph by disturbing Reissner's membrane. He made very small iron balls, about 0.05 mm in diameter, and through a hole in the bony labyrinth of a guinea pig introduced one onto Reissner's membrane, arranged so that the particular portion was flat, see Fig. 5.26. With an electromagnet he caused the ball to be suddenly lifted from the membrane and followed the potential change in the perilymph of the scala vestibuli

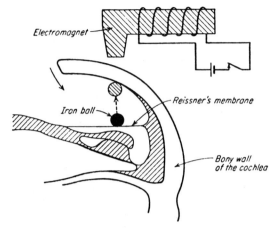

FIG. 5.26   Schematic of method of imparting a mechanical shock to Reissner's membrane. [From von Békésy (1951a).]

with the electrode connected to an oscilloscope. The trace is shown in Fig. 5.27. It is seen that there is a sudden rise in potential with the lifting of the ball and that the subsequent natural oscillation of the membrane, about 150 Hz, is damped in about three or four oscillations. He then deprived the animal of oxygen and found that as death was approached by anoxia, the same waveform was obtained and, by increasing the gain of the oscilloscope, he could duplicate the original trace.

Although the above experiments showed that a displacement of the basilar membrane resulted in a cochlear microphonic, they did not determine if the microphonic arose from the displacement of the membrane or

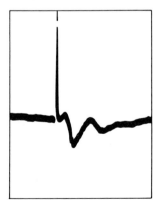

FIG. 5.27   Cochlear microphonic when iron ball of Fig. 5.26 was suddenly lifted by the electromagnet. [From von Békésy (1951a).]

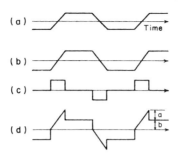

FIG. 5.28 (a) Trapezoidal vibration applied to basilar membrance. (b) Expected voltage response of choclear microphonic is proportional to displacement only. (c) Expected voltage response if proportional to velocity of displacement only. (d) Expected voltage response if proportional to both displacement and velocity. [From von Békésy (1951b).]

the velocity of its displacement. von Békésy also resolved this problem by another ingenious technique. He developed a vibrating needle which he could place on the basilar membrane while controlling the amplitude and frequency of the vibration. The microphonic output was displayed on an oscilloscope from the usual electrode in the scala vestibuli. von Békésy altered the waveform of the vibrating needle from sinusoidal to trapezoidal by controlling the waveform of the input current. Consider the waveform of the needle displacement shown in Fig. 5.28a. If the cochlea microphonic is proportional only to the displacement of the basilar membrane, the waveform of the microphonic will be the same as that of the displacement, Fig. 5.28b. If, however, the microphonic is proportional to the velocity of the displacement, its waveform for a trapezoidal displacement will be of the form of Fig. 5.28c. If there is a proportionality to both velocity and displacement, the microphonic will have the form of Fig. 5.28d. The experimental result is shown by the oscilloscope trace of Fig. 5.29. Clearly, the microphonic is proportional only to the displacement of the basilar membrane. A decrease of oxygen supply to the animal resulted in a rapid decay of the plateau of the trapezoid during oscillation.

A further experiment by Békésy showed that the endolymph in the scala media undergoes a larger potential change than does the perilymph during

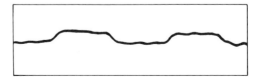

FIG. 5.29 Oscillograph of the microphonic response to the trapezoidal stimulus shown in Fig. 5.28a. [From von Békésy (1951b).]

a cochlear microphonic and that the deformation of Reissner's membrane contributes very little compared to that of the basilar membrane. A small capillary was inserted through a hole in the bony wall of the cochlea just above Reissner's membrane and a small amount of perilymph was drawn into the capillary. The vibrating needle was placed in contact with the fluid in the capillary tube and an electrode placed in the perilymph. If this fluid-capillary pulsator were located outside of Reissner's membrane, Fig. 5.30a, Reissner's membrane and the basilar membrane would move in phase, i.e., for movement of the capillary fluid downward both membranes would move downward. If, however, the capillary penetrated Reissner's membrane, the two membranes would move out of phase with one another, i.e., for injection of fluid from the capillary, the increased hydrostatic pressure in the scala media would force both membranes outward, Fig. 5.30b. Consequently, if Reissner's membrane was responsible for the potential, the phase of the microphonic would reverse, i.e., be out of phase with the vibrator, when the membrane was penetrated. The experiment showed that there was no phase reversal of the microphonic when Reissner's membrane was penetrated. This result showed that the main source of the microphonics is associated with the basilar membrane. This conclusion was confirmed by pushing the capillary in further until it penetrated through the basilar membrane. As seen in Fig. 5.30, such a penetration will reverse the motion of the basilar membrane, and, indeed, the phase of the microphonic is reversed with respect to that of the vibrator.

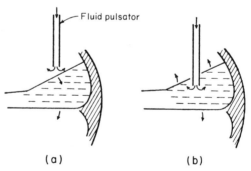

FIG. 5.30 Stimuli with fluid pulsator (a) above and (b) below Reissner's membrane. [From von Békésy (1960).]

The observation that the cochlear microphonic is functionally dependent only on the displacement of the basilar membrane permitted more detailed experiments. Tasaki *et al.* (1954) measured the magnitude of the microphonic in the endolymph of the scala media when the hydrostatic pressure in each of the three scalae was changed. This was done with an electrode in the endolymph and micropipettes in the three scalae. The dc potential in the endolymph of the scala media increased when positive pressure was applied to the scala vestibuli; it decreased when positive pressure was applied to the scala tympani; and it increased when positive pressure was applied to the scala media. Negative pressures in the three scalae were obtained by suction on the pipettes. In summary, a reduction in the positive dc potential of the endolymph is caused by

(1) negative pressure applied to the scala vestibuli,
(2) negative pressure applied to the scala media, and
(3) positive pressure applied to the scala tympani.

Note that the displacement of the basilar membrane due to these three maneuvers is upward, i.e., toward Reissner's membrane. Up to a limit, the magnitude of the potential change is correlated with the magnitude of the displacement of the basilar membrane.

The above experiments show that the cochlear microphonic can be measured by placing an electrode in any of the three scalae if one is not concerned with the phase of the output. The ground or indifferent electrode is placed somewhere on the body of the animal, usually the neck. Since external sound also causes displacement of the basilar membrane, the relative magnitudes of the cochlear microphonic potential can be measured by the peak-to-peak voltage of the microphonic produced by a pure sine wave of external sound of constant frequencies and acoustical pressure. In this way, the frequency response and pressure response of each of the four turns of the cochlea can be compared, i.e., along the basilar membrane from the window to the helicotrema. (Note that although there are only 2.5 loops in the cochlea, Fig. 5.3, if one drills holes downward, or at some other angle, four openings at different "turns" may be made, counting ones near the window and the apex.)

An example of typical data for the basal turn (nearest the window), turn II and turn IV (apical), is shown in Fig. 5.31. A shift in magnitude of peak-to-peak voltage of an electrode in the perilymph is seen to occur with frequency. This effect has been studied in detail by Tasaki et al. (1952). Example data by these investigators are shown in Figs. 5.32a and b, a for turn II and b for turn III. The curves measure the peak-to-peak voltage in the perilymph versus the sound pressure. Note that since the plots are log–log, the straight-line parts of the curves have a slope of unity, which implies a linear relationship between cochlear potential and pressure. The behavior of turns II and III shows that as the distance along the basilar membrane increases, the linear response for high frequencies is suppressed. This agrees with the mechanical behavior of the traveling wave discussed earlier. That is, since the basilar membrane is smallest near the round window, it is most responsive to high frequency sound in that region. As it

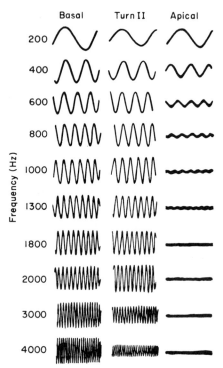

FIG. 5.31 Cochlear microphonics in guinea pig at oval window, turn II and turn IV (apex) for different frequencies. Intensity adjusted for constant response in basal turn; paired electrodes, scalae vestibuli and typmani in each turn. [From Tasaki *et al.* (1952).]

grows wider toward the helicotrema, it is more responsive to low frequencies. The data taken by Tasaki *et al.* can be plotted in another way, that is, the cochlear microphonic voltage at each of the four turns versus frequency of the input tone. The sound pressure level was adjusted at each frequency to keep the voltage output constant at turn I. Two different intensity levels were used, and these are shown as the short horizontal lines in Fig. 5.33. We see in this figure essentially a plateau for turns I and II, followed by an exponential decrease in output. The decrease occurs at about 10 kHz for turn I and 2.5 kHz for turn II. Turns III and IV show no plateau, only the exponential decrease. As expected, for the heavier end of the basilar membrane there is no response to high frequencies.

SUMMATING POTENTIAL

When an electrode is placed in the cochlea with the indifferent electrode inserted in a wound in the neck, a cochlear microphonic potential is

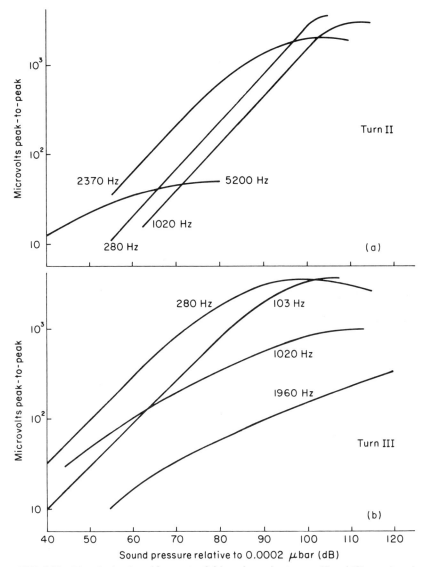

FIG. 5.32   Magnitude of cochlear potential in guinea pig at turns II and III as a function
of sound pressure at different frequencies. [Redrawn from Tasaki *et al.* (1952).]

observed when a sound stimulus arrives at the oval window of the ear.
Generally, this microphonic follows the impressed sound wave (see Fig.
5.31) but is too large for details to be observed. In a series of careful
experiments, Davis *et al.* (1958) used a differential amplifier with which
they were able to observe details within the microphonic. When the

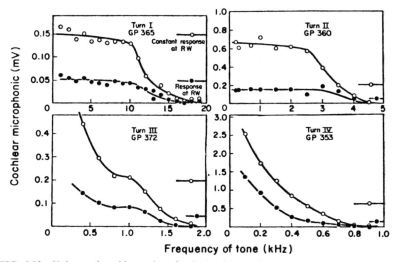

FIG. 5.33 Voltage of cochlear microphonic (peak-to-peak) as a function of frequency at the four turns of the cochlea. Four different guinea pigs were used (GP). The upper and lower curves represent two different intensities of sound. [From Tasaki *et al.* (1952).]

positive pressure phase (condensation) of a sound wave strikes the oval window, the microphonics of the scala vestibuli and the scala media become positive in potential, while that of the scala tympani becomes negative. The opposite potentials occur for the negative part of the sound wave. From other experiments, they also observed that an electrode placed against the round window of the scala tympani recorded the same potential changes as one inserted in the scala tympani. When they added the output of an electrode in the scala vestibuli to that of one in the scala tympani, subtraction of the microphonic potentials was obtained since they are opposite in polarity. By this method they could study the remainder. Two components of the remainder were observed and studied in detail: the action potential and the summating potential. The action potential was observable because the peripheral terminals of the fibers of the auditory nerves are closely connected to the source of generation of the cochlear microphonic, and their signals are picked up by an electrode in the cochlea. The differential amplifier that they used also permitted the determination of the phase differences of the signals. Furthermore, by control of the amplification of the two channels, complete cancellation of the microphonic could be obtained.

Electrodes were placed in the scalae of a guinea pig, as shown in Fig. 5.34. The cochlear microphonic which follows the course of the tone pip, observed from an electrode at the round window, is shown in the upper (oval) part of Fig. 5.35. The time course of this trace smears the individual

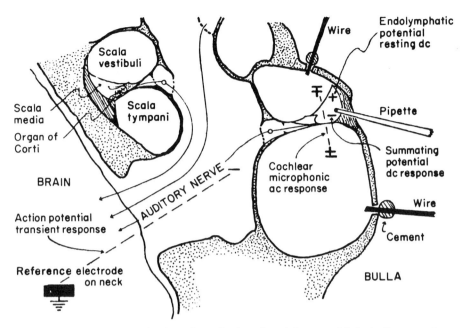

FIG. 5.34 Position of the electrodes and orientation of the potentials in the first turn of a guinea pig cochlea. [From Davis *et al.* (1950).]

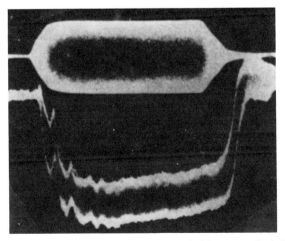

FIG. 5.35 Upper part is cochlear microphonic recorded from round window of a guinea pig in response to a strong tone burst of 21.5 kHz. Lower curves are summating potentials with action potentials in the early part. Upward deflection means scala vertibuli is more positive relative to scala tympani or else cochlea is more positive relative to neck. [From Pestalozza and Davis (1956).]

parts of the 21.5 kHz wave, so that all that is seen are the average positive and negative parts of the microphonic. When the outputs from two electrodes are added, the cochlear microphonic potential is zero and the remainder is shown in the two lower scope traces. The peaks in the traces at early times are the action potentials. The long depressed signal is the summating potential; the upward deflection on the oscilloscope means that the scala vestibuli is more positive relative to the scala tympani. Thus, the summating potential appears as a shift in the baseline of the cochlear microphonic potential, upon which the action potential spikes can be seen.

The important point is that the magnitude of the summating potential is frequency dependent. It is not detectable until about 2000 Hz, but at 8000 Hz it dominates. The relative threshold of detectability for a guinea pig, which can hear much higher frequencies than humans, is shown in Fig. 5.36. At low frequency the cochlear microphonic is in phase with each individual cycle of an audio signal, while at high frequencies the cochlear microphonic no longer responds in phase but is taken over by an averaging of several audio cycles and appears as an average potential shift, hence the name summating potential.

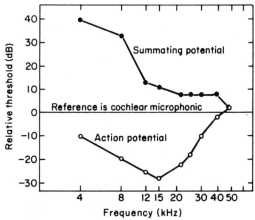

FIG. 5.36 Summating and action potential threshold versus frequency. [From Pestalozza and Davis (1956).]

ACTION POTENTIALS

The action potential characteristic of single fiber nerve conduction has been treated in detail in Vol. I, Chapter 3, and has been shown to be the mode of signal conduction to the brain for body sensors in the preceding chapters. A number of studies of the action potential from individual nerve fibers in the auditory system have been carried out, and we will describe some of the findings of one of the earlier investigators. Tasaki (1954) performed experiments on guinea pigs with microelectrodes ($\frac{1}{4}$- to $\frac{1}{3}$-μm diameter) inserted into the nerve bundle leaving the cochlea. The length of time in which good data could be taken from a single nerve ranged from 10 to 100 sec, so detailed information was difficult to obtain. However, he was able to elucidate the general principles.

The sound source was short pips of pure tones with both frequency and intensity as variables. The response of a single fiber was somewhat random in that it would only sometimes yield action potentials from a stimulating sound pip.

When a sinusoidal sound wave strikes the oval window, the potential of the cochlear microphonic follows the cycle, although the phase relation of the microphonic changes with distance from the window. In addition, we have seen that the response at different distances from the window is frequency dependent. These two phenomena are shown in Fig. 5.37. When the sound wave is displayed on an oscilloscope simultaneously with the action potentials from single fibers, it is seen that the nerve impulses start at a definite phase of the microphonic wave. This can be seen in Fig. 5.38. In this figure the lower curve is the sound wave, which is also the shape of the microphonic potential, and the upper curve is the corresponding action potential. This relationship of the action potential to the phase of the microphonic had been observed earlier in whole nerves by Rosenblith and Rosenzweig (1952).

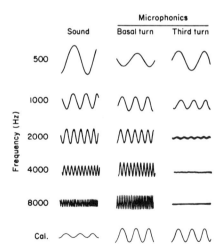

FIG. 5.37   Relation between sound intensity at different frequencies and microphonics at turns I and III in guinea pig recorded simultaneously. [From Tasaki (1954).]

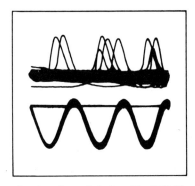

FIG. 5.38 Single fiber spikes in guinea pig induced by 1000 Hz pure tone. Sound intensity at 55 dB above human threshold. Sinusoidal wave (lower) recorded with microphone. Note that action potentials (upper) occur at a particular phase of sound wave. [From Tasaki (1954).]

By exploring the responses of a large number of single fibers, Tasaki was able to show that there is a frequency–intensity relation for each. A typical example of this is shown in Fig. 5.39 for a single fiber. (Note that the ordinate is not defined in the original paper, so the absolute values are not known.) At the highest intensity, 0 dB, this particular fiber will respond to sound pips up to 7.5 kHz, while at the lowest intensity the range of response is severely restricted. The dashed line delineates the boundary of

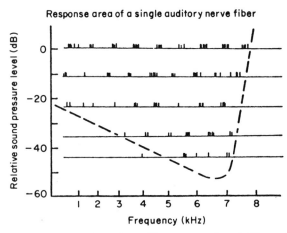

FIG. 5.39 Responses of a single auditory fiber to tone pips of different frequencies and intensities. Dotted line shows boundary of response area of fiber. [From Tasaki (1954).]

this "response area" for this particular fiber. The extreme steepness of the
boundary of the response area on the high frequency side indicates that at
a given position in the cochlea, the mechanical vibration becomes suddenly
very small when the frequency is increased above a certain limit. This
observation is in agreement with the mechanical measurements of von
Békésy (Fig. 5.13).

Tasaki further showed a general relationship for the variation of the
number of impulses in a single fiber with sound intensity. A typical set of
data is shown in Fig. 5.40. The solid line in the figure was drawn in
accordance with the formula

$$N = D(1/k)\ln(I/I_0) \tag{5.7}$$

where N is the number of pulses induced by a sound pip of duration D and
intensity I, and k and I_0 are constants chosen to fit the curve. This is a
form of the Weber–Fechner Law which will be discussed in Chapter 7.

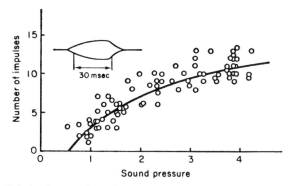

FIG. 5.40   Relation between intensity of a pip and number of impulses (spikes) elicited in
a single auditor fiber. Pips were 5000 Hz and insert shows shape of pip. Sound pressure is in
microbars. [From Tasaki (1954).]

THE AUDITORY PATHWAY

The precise delineation of the pathway of a tonal stimulation of the
organ of Corti to the cerebral cortex cannot, at present, be determined.
Such a delineation implies a classical interpretation of the form developed
by Ramon ý Cahal that would include fibers amd nuclei through which an
anatomical tracing of a series of sequential connections could be made.
Such is not the case. The tracing of the pathway can be seen in Fig. 5.41.

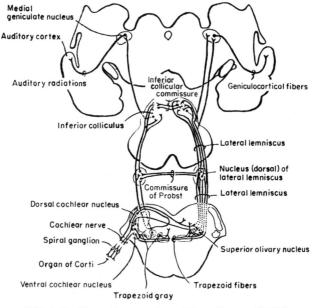

FIG. 5.41   The auditory pathway. [From Guyton (1971).]

Without delving into the detailed physiology of the pathway, the general principles may be stated (Guyton, 1971; Ades, 1959; Galambos, 1954).

(1)   The fibers leave the organ of Corti and enter into the upper part of the medulla.

(2)   At this point the fibers synapse with second-order neurons.

(3)   Some of these second-order neurons pass to the opposite side of the brain, while some remain on the same side and pass to the *superior olivary nuclei*.

(4)   Most of the fibers entering the superior olivary nuclei terminate there, but some pass through.

(5)   In their upward course, some fibers cross to the opposite side of the brain through the *commissure of Probst*, while still others cross through the *inferior collicular commissure*.

(6)   The pathway then extends 'to the *medial geniculate nucleus* where cell fibers synapse.

(7)   From here, the auditory tract spreads by way of the *auditory radiation* to the *auditory cortex* located mainly in the superior *temporal gyrus*.

From this summary it should be noted that impulses from either ear are transmitted through both sides of the brain, and there is a partial crossing over between the two sides in at least three different places. Also, the tract consists of at least four types of neurons which may or may not synapse in the varying locations. This means that some of the pathways are more direct than others and, accordingly, some impulses may arrive at the cortex ahead of others. Two areas of the cortex respond to long and short latency, respectively, Fig. 5.42.

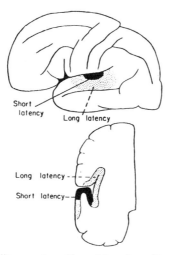

FIG. 5.42   The auditory cortex, side and top views. [From Guyton (1971).]

Very little is known about the transmission of auditory signals through this pathway. There have been some investigations of cells in the cortex of animals and their response to auditory signals. Usually, an exploring electrode is inserted in the auditory cortex of an anesthetized animal and sound clicks or tone signals are made near the ear. Such data are difficult to interpret because the responses are dependent upon the depth of the anesthetization.

It will be shown in Chapter 6 that there is almost a point-to-point mapping of the retina of the eye on the striate cortex. There is no such precise mapping of sounds in the auditory cortex. Some neurons have been located which are frequency selective at a low intensity of sound, but this selection may broaden as the intensity increases. Some of these neurons fire spontaneously and cease when stimulated by sound of the proper frequency. Examples of recordings from such cells are shown in Fig. 5.43. Traces in Figs. 5.43(a) and (b) show that the spontaneous firing of an auditory cortex cell ceases when the ear is stimulated by a 2-kHz tonal frequency. Trace (c) shows a cell type which, while normally inactive, fires when tonal pips of 1 kHz are used as a stimulus. Traces in Figs. 5.43 (d)

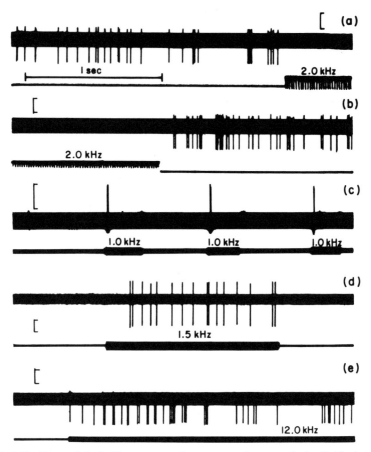

FIG. 5.43 Types of single fiber responses in a cat to auditory tonal stimuli. The height of the bracket indicates 100 μv. Lower traces represent onset and termination of stimuli: (a) and (b) A spontaneous discharging unit is suppressed by a 2-kHz tone. (c) Discharges occur only at onset of 1-kHz tone. (d) Normally off fiber discharges during 1.5-kHz tone. (e) Normally off fiber discharges during 12-kHz tone. [Reprinted with permission from Erulkar, Rose and Davies, *Bull. Johns Hopkins Hosp.* **99**, 55 (1956). Copyright © 1956 by The Johns Hopkins University Press.]

and (e) show sustained discharges of two different cells during tonal stimulation by 1.5 and 12 kHz, respectively.

Referring back to Fig. 5.39, it is seen that an auditory fiber can have its frequency sensitivity broadened by an increase in intensity, but always the broadening is toward the lower frequencies. Thus, it may be said that each fiber will respond up to its highest frequency but to none higher. It is apparent, then, that frequency discrimination by the ear is not a process of the vibrational pattern within the cochlea alone, but must take place later in the auditory pathway. Removal of the auditory cortex in animals trained to respond to a series of tones does not prevent the animal from detecting and responding to sounds, but it does remove his trained response to the tonal pattern. Therefore, frequency discrimination takes place within the auditory cortex by some complex arrangement of excitation and inhibition of the neurons.

THE SOURCE OF THE COCHLEAR MICROPHONIC

We have shown that there is a resting potential difference between the scala media and the other scalae and that this potential changes with displacement of the basilar membrane. And there is also a potential difference along the scala vestibuli, as shown in Fig. 5.25. von Békésy (1960) was able to demonstrate that the cochlear microphonic could not arise from a transducer action such as that in a piezoelectric crystal (Vol. I, Chapter 4).

Consider his findings. The experiment in which he stimulated the basilar membrane with a trapezoidal impulse, Fig. 5.28, showed that the cochlear microphonic responds only to displacement. This precludes any theories based on potential differences produced during movements of fluids along a solid wall; trapezoidal stimulation of these friction potentials would produce the velocity dependent signal of Fig. 5.28c, a form that is not observed.

The theoretical possibilities for the origin of the microphonic are further limited when the actual energy dissipation measured in the cochlea is compared to the acoustic energy. From Fig. 5.25, it is seen that if two electrodes are placed in the scala vestibuli a potential difference is measured, for example, between positions 1 and 3 of Fig. 5.44. The resistance of the vestibular canal can be determined by measuring the change in resistance when the electrode at position 3 is moved to position 2. The resistance of the scala vestibuli from the round window to the helicotrema is about 15,000 Ω. If E_0 is the potential difference, measured as 2 mV, between these two points, there is a continuous energy loss in the scala vestibuli of E_0^2/R. If one of the electrodes is pushed against the Reissner's membrane, the potential of the perilymph near the helicotrema increases relative to that near the oval window. Call this potential increase E; the new potential difference will be $E_0 + E$ and the new rate of energy loss will be $(E_0 + E)^2/R$. The rate of energy lost per second due to this displacement is

$$W = \frac{(E_0 + E)^2}{R} - \frac{E_0^2}{R} = \frac{2E_0E + E^2}{R} \quad \frac{\text{W}}{\text{sec}} \qquad (5.8)$$

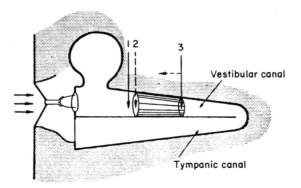

Vestibular canal

Tympanic canal

FIG. 5.44 Longitudinal section of straightened cochlea. [From von Békésy (1962).]

[Note that in von Békésy (1960) this equation is printed incorrectly.] In an actual experiment von Békésy found that a force on a small probe required to displace Reissner's membrane by 10^{-5} m was 1.4×10^{-5} N. The work of depressing the membrane, assuming it obeys Hooke's law, is

$$W = \int_0^x F\,dx = \int_0^x kx\,dx = \tfrac{1}{2}kx^2 \qquad (5.9)$$

and, from $k = F/x$, $k = 1.4 \times 10^{-5}/10^{-5} = 1.4$ N/m. Therefore, the work of displacement is

$$W = \tfrac{1}{2} \times 1.4 \times \left(10^{-5}\right)^2 = 7 \times 10^{-11}\ \text{J} = 7 \times 10^{-11}\ \text{W/sec}$$

The increase in potential due to this displacement was measured to be $E = 0.3$ mV, and therefore the electrical energy loss per second is

$$W = \left[2 \times 2 \times 10^{-3} \times 0.3 \times 10^{-3} - \left(0.3 \times 10^{-3}\right)^2\right]/(15 \times 10^3)$$
$$= 8.6 \times 10^{-11}\ \text{W/sec}$$

In 1 sec, the electrical energy becomes greater than the mechanical work done on the membrane, and Békésy found no difficulty in maintaining this current for 10 sec or more. It can therefore be concluded that the system which produces the microphonics is that of an amplifier rather than a transducer that simply converts mechanical to electrical energy.

Following this pioneering work of von Békésy, a search for the biological battery was conducted by Tasaki et al. (1954). By carefully inserting 0.3-μm diameter electrical probes and taking the potential in the scala tympani as the reference point, they produced the potential map of Fig. 5.45. During acoustic stimulation, the largest ac microphonic was found to be in the scala media. The voltage is less, but in the same phase, in the scala vestibuli and still less in the spiral ligament. The voltage is also strong

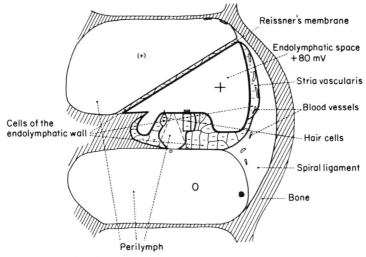

FIG. 5.45   Distribution of resting potentials in the guinea pig. [From Tasaki *et al.* (1954).]

in the stria vascularis and nearly in phase with the scala media. It is very strong in the region of the hair cells but is opposite in phase to that in the scala media. A schematic of the voltage changes, while the ear was stimulated by a 500-Hz tone, recorded by a microelectrode advanced from the scala tympani to the scala media is shown in Fig. 5.46. The cochlear microphonic is shown at the left and the path of the electrode is upward from the bottom (in the direction of the dashed arrow). It is seen that at the reticular lamina in which the hair cells are embedded (see Fig. 5.18),

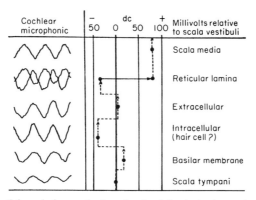

FIG. 5.46   Potentials and phases of microphonic while electrode was inserted in direction from bottom to top of figure during constant stimulation with 500 Hz. [From Tasaki *et al.* (1954).]

there is a shift in phase of the microphonic coinciding with the dc change from −35 to +80 mV. The path of the ac current proposed by Tasaki, Davis, and Eldredge is from the scala media across Reissner's membrane, mostly by capacitive conductance, to the scala vestibuli and then to the scala tympani. There may also be significant capacitive current flow across the stria vascularis to the spiral ligament. The stria vascularis was assumed to be the most likely source of the biological battery because the highest dc potential was in the endolymph.

The amplification process proposed by Davis (1965) is illustrated in Fig. 5.47. The biological battery which maintains the dc potential in the scala media is shown in the stria vascularis. When the basilar membrane vibrates, a shearing motion occurs between the reticular lamina and the tectorial membrane which causes the hairs of the hair cells to bend. This bending causes a change in the electrical resistance across the membrane at the base of the hairs, and the leakage current will either increase or decrease. This change of current can either liberate a chemical mediator which stimulates the nerve ending (synaptic transmission) or else the current directly excites the nerve ending (ephaptic transmission). In this diagram only the circuit for the outer hair cells is shown. This postulated circuit is that of an electronic amplifier where the hair cell motion behaves

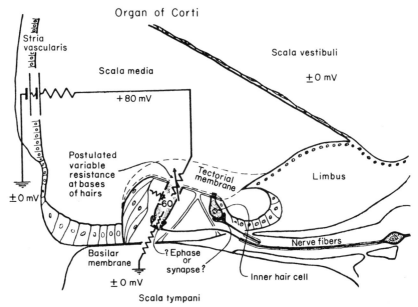

FIG. 5.47   Resistance microphone theory of electrical amplification. The primary battery is in the stria vascularis. Leakage current flows through the hair cells whose resistance is altered by bending. [From Davis (1961).]

similarly to the grid of a vacuum tube or a junction potential of a semiconductor.

ACOUSTICAL QUANTA AND THE THEORY OF HEARING

The development of quantum mechanics was based on the de Broglie hypothesis that all particles could be considered to have wavelike character with a wavelength given by the relation

$$\lambda = h/p$$

where p is the momentum mv and h is Planck's constant. The wave nature of Newtonian particles was demonstrated by Davisson and Germer who showed that electrons could be diffracted from crystals, with appropriately spaced atoms, as if the electrons were waves with a wavelength given by the above de Broglie relation. Because of this relationship, the mathematics of wave behavior, developed in.the nineteenth century, were promptly adapted to the new quantum, or wave, theory of matter. Acoustical waves were used as illustrations of the wavelike behavior of particles .

Concurrent with the early development of quantum mechanics, Heisenberg postulated his now famous "uncertainty principle"; the product of the uncertainty of simultaneous measurements of both the position and momentum of a particle is equal to or greater than Planck's constant,

$$\Delta x \, \Delta p \geqslant h \qquad\qquad (5.10)$$

The reason for this is that any measurement of either quantity requires the obtaining of information by either the emission of a quantum of energy or the reflection of one from an external source. Each effects a change which alters either the position or momentum of the particle; an attempt to increase the accuracy of one of the parameters decreases the precision of the other.

There is another useful form of the uncertainty principle. Suppose we wish to measure the energy E emitted during a time interval Δt in some atomic process. If the emission is an electromagnetic wave, there is a limited time available to determine its frequency. The absolute minimum in time is that required for a single cycle of the wave, i.e., for less than a single cycle the wavelength, and hence the frequency f, could not be determined with accuracy. The number of waves divided by a time interval Δt is the definition of frequency and therefore the uncertainty in the measurement of a frequency Δf is

$$\Delta f = 1/\Delta t$$

The energy of a photon is given by

$$E = hf$$

and the uncertainty in the energy is proportional to the uncertainty in the frequency or

$$\Delta E = h\,\Delta f$$

therefore

$$\Delta E = h/\Delta t \cdot$$

or

$$\Delta E\,\Delta t \geqslant h \tag{5.11}$$

where the equal to or greater sign is introduced because, as mentioned above, the minimum time for accurate determination of frequency is 1 cycle. Note that if one substitutes $\Delta E = h\,\Delta f$, the relation may be written as

$$\Delta f\,\Delta t \geqslant 1 \tag{5.12}$$

Stewart (1931) first used the uncertainty principle of quantum mechanics to develop an equivalent one for acoustical waves and thereby developed the concept of an acoustical quantum. Gabor (1946) extended this concept to the analysis of some key experiments. An understanding of this analysis requires a knowledge of mathematical techniques beyond the scope of this book. However, its implications are so important for the theory of hearing that a brief description will be given. The reader interested in learning more of this subject should perhaps begin with Gabor (1947).

We have discussed in this chapter two types of sounds used in experiments, a pure tone and a click. A pure tone is a sinusoidal wave of a single frequency f which can last for any time Δt. The click is made up of a mixture of almost all audible frequencies with a very short duration. Such a mixture of frequencies is expressed as a sum of Fourier components, that is, the product of the amplitude of each frequency and its corresponding sinusoidal behavior with time. Such a spectrum of frequencies can be considered as a distribution about a mean, and its effective width in frequency being that within some high and low cutoff values (since a Fourier series is infinite). Gabor then constructed a schematic graph with time as the ordinate and freqnency as the abscissa, Fig. 5.48. On the left of the ordinate is a typical sinusoidal wave, and it should be recalled from Vol. I, p. 207 that the square of such a wave is proportional to its energy. This square, which is only positive, is shown by the shaded curve to the left of the ordinate and the sum of the shaded areas is the energy of that part of the pure sinusoidal wave. If the wave is continued only for a time Δt, its effective duration, it contains the energy shown in the figure. Below the abscissa is the real part of the Fourier spectrum of the click. The sum of the squares of its components is shown by the shaded curve below the abscissa. Its effective width in frequency is somewhat arbitrarily taken as the width at half-maximum on either side of the mean frequency as shown. This is chosen to give the same total amount of energy in its signal as that of the Δt length of the pure tone. (Actually, effective duration Δt and the

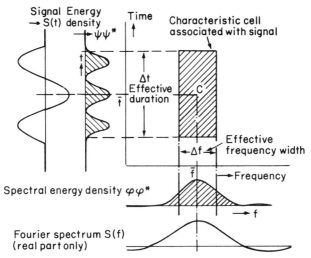

FIG. 5.48   Schematic drawing in which the area of the rectangle is the product $\Delta f \, \Delta t$. On the far left is a pure sinusoidal sound wave of frequency f and duration Δt, the shaded area is its square or energy density. At the bottom is a sound with a frequency spectrum above a mean value f, with the shaded curve its square or energy density. The section within the range Δf has the same total energy as the pure tone. [Reprinted by permission from Gabor, *Nature* **159**, 591 (1947). Copyright © 1947 Macmillan Journals Limited.]

effective frequency width Δf are defined by the mean square deviations from the mean values of t and f, as indicated.) Gabor then showed mathematically that

$$\Delta t \, \Delta f \geqslant 1$$

by appropriate normalization of time and frequency behavior. Thus, the shaded rectangle has an area of at least unity. This is the same formulation as the uncertainty principle of quantum mechanics, Eq. (5.12), and this classical analog may be called an acoustical quantum.

Concurrent with Gabor's analysis, an experimental method was developed at Bell Laboratories on "sound spectrography" by Potter (1945) and Kopp and Green (1946) and described in the book by Potter *et al.* (1947). These investigators used a series of tuned resonators whereby speech could be displayed on a time versus frequency chart. These experiments agreed with Gabor's analysis. A discussion of this work is too lengthy for our purposes. We will characterize instead three other types of experiments and indicate from them how Gabor was able to demonstrate that the ear possesses a threshold area of discrimination, $\Delta f \, \Delta t \geqslant 1$, and how this leads to the conclusion that the ear possesses a minimum of two hearing mechanisms.

(1) If a pure sinusoidal oscillation is sounded for a few cycles only, the ear will hear it as noise. From 10 msec or longer duration, the ear will hear

it as a short musical note with an identifiable frequency (Bürke *et al.*, 1935).

(2)   A single frequency was sounded for a short time and then doubled in intensity. If the time was less than 21 msec (at 500 Hz), the step of the sound could not be differentiated and the same sensation was obtained as if the sound had the double intensity from the beginning. It appears that in the interval between 10 and 21 msec the ear has not quite become ready to register a second distinct sensation (Bürke *et al.*, 1935).

(3)   When a pure note is sounded, say at 500 Hz, the minimum frequency change for which a small oscillation from 500 Hz can give the sensation of a "trill" is about 2.3 Hz. The ear cannot distinguish a frequency shift from the pure note with changes less than this. The shift time which gives the finest frequency discrimination is 0.5 sec, or a time of 0.25 for one up or down shift (Shower and Biddulph, 1931).

In each of these cases there is a time and frequency interval and it is possible to form a product, the "threshold area," in which the ear is capable of registering one sensation only and which must be exceeded if it is to register a second sensation. Using the original data of these experiments, Gabor was able to calculate a time–frequency discrimination product and show that it is essentially of the order of unity, or the "acoustical quantum limit," Eq. (5.12).

As a rough example, the experiment of item (3) above indicated that the ear had its finest frequency discrimination when the frequency swing time was 0.5 sec and the amount of frequency shift required was 2.3 Hz, giving a product of about unity. If, on the other hand, a short sound with poorly defined frequency strikes the ear, the time discrimination, as indicated in (2) above, is about 20 msec. And, as the experiments of (3) show, if the sound is prolonged, the bandwidth of the frequency discrimination shrinks to only a few cycles, while the ear keeps in step with the sound for at least 1/4 sec (half or a 0.5 sec full cycle).

The importance of these findings is related to the mechanism of hearing. An older hypothesis was that the ear had finely tuned resonators at each frequency. No such system of resonators could yield the above experimental results which verify Eq. (5.12). That is, resonators cannot change their frequency sensitivity if the time of the signal is varied. Therefore, a second mechanism in the ear must be postulated which accounts for its progressively refined tuning ability as the duration of the sound is prolonged.

Gabor proposed that the requirements of two mechanisms of hearing indicate that there is (a) a mechanical one of finely tuned resonators, and (b) a nonmechanical one, which after a certain time selects the resonator at the maximum of the frequency spectrum being perceived. The behavior of the two mechanisms is illustrated in Fig. 5.49, in which time of a sound vibration is plotted against frequency. The broad curve is symbolic of the

disturbance of the basilar membrane and the resulting apparent broad
resonator reception of the spectrum of the sound if it is for less than 10
msec, as pointed out in (1) on p. 143. As the length of time the sound is
heard increases, a second mechanism starts searching for the maximum of
excitation and locates it with an accuracy of about 3 Hz by 250 msec in a
way in which the product of the time (up to 250 msec) and the frequency
discrimination width remains a constant of the order of unity.

Gabor suggested that the frequency discrimination "mechanism" need
not be assigned to the brain. Instead, there might be a new phenomenon in
nerve conduction. If conduction of adjacent nerve fibers in a bundle is
somewhat unstable, the one that conducts a signal most strongly, i.e., the
one at the frequency maximum in a spectrum, may gradually suppress the
conduction of its neighbors. Thus, if a pure note is sounded for a sufficient
time, less and less fibers will be excited until only the one corresponding to
the position of maximum amplitude in the basilar membrane will conduct.
This mechanism would make it possible to distinguish notes accurately in
spite of the broad frequency spectrum of oscillation of the basilar membrane, Fig. 5.13.

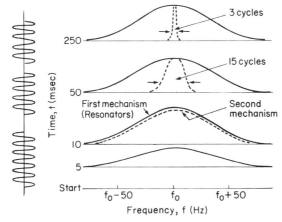

FIG. 5.49   Schematic drawing illustrating the two mechanisms of hearing. The mechanical
mechanisms of resonators dominates over a broad frequency range up to about 20 msec. For
longer times a neural mechanism locates the maximum and suppresses the other frequencies.
[Reprinted by permission from Gabor, *Nature* **159**, 591 (1947). Copyright © 1947 MacMillan
Journals Limited.]

HEARING DEFECTS

There are two basic types of deafness: conductive deafness which arises
from impairment of the mechanical system of the ear and neural deafness
which arises from impairments of the electrophysiological systems of the
inner ear and nervous pathways. Simpler types of conductive deafness are

readily determined, such as blockage of the external ear canal or performation of the eardrum. Other mechanical impairment can arise from calcification of the ossicles or a breakage of one of the ossicles. Inner ear infections or perforation of the tympani membrane also result in mechanical impairment. A simple test for mechanical impairment relies on the fact that there is no bone conduction impairment. Thus, a tuning fork held near the ear may not be heard loudly, but it will be if it is placed against the mastoid bone behind the ear. Conductive hearing loss can be fully compensated for by an increase in loudness, such as turning up the volume of a radio of using a hearing aid. The individual with such impairment is somewhat handicapped because loud speech, in the absence of an electronic aid, is rarely articulated well, so that there is some loss in discrimination. Nerve deafness, on the other hand, shows no change in hearing ability between bone and air conduction. It tends to reduce the high frequencies to a greater extent than low frequencies, although this is not always the case. Modern hearing aids, miniturized with semiconducting modules, have tone as well as loudness controls, much like a radio. Thus, some control over the relative intensity of the frequency spectrum can be attained.

THE FENESTRATION OPERATION

Otosclerosis is a disease of the bone in the region of the oval window that leads to partial or complete fixation of the stapes to the window. Tests for this condition and the improvement by a fenestration (window) operations are carefully reviewed by Wever and Lawrence (1954) and some selections from their coverage will be summarized.

The fixation of the oval window not only eliminates the hydraulic transformer action of the ossicles but it also immobilizes the cochlear fluid. Thus, any sound waves on either window can cause no motion of the basilar membrane and therefore no stimulation of the organ of Corti.

Early attempts were made to loosen the stapes or remove it. This operation did not have a lasting effect because otosclerosis soon closed the opening or a continual loss of perilymph caused deafness. Furthermore, the operation was criticized as dangerous. Pioneering surgeons began to

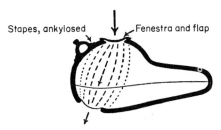

FIG. 5.50 Schematic of the motion of a sound wave in a fenestrated ear. [From Wever and Lawrence (1954).]

make an opening in one of the semicircular canals of the vestibular labyrinths, but the closure with ear tissue soon deteriorated. The credit for making the fenestration procedure into a practical operation belongs to Julius Lempert.

Figure 5.4 shows how the semicircular canals are continuous with the scala vestibuli and tympani, and they all contain the same perilymph. If an opening is made in one of the canals and covered with a flap of tissue, motion of the basilar membrane can take place. This is shown schematically in Fig. 5.50.

When the inner ear is opened, the three bulges of the bony walls of the three semicircular canals are clearly seen, Fig. 5.51. The incus and the head

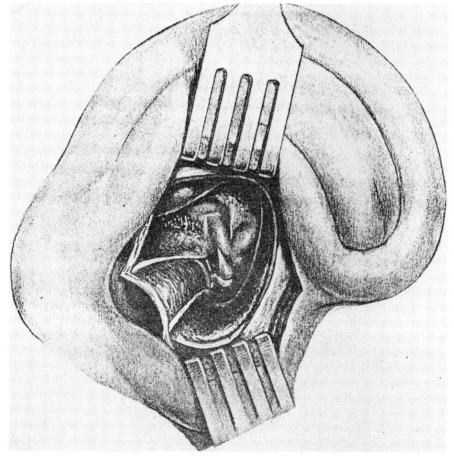

FIG. 5.51 Exposure of the ear in Lempert fenestration operation. The malleus and incus are visible but the stapes is in shadow. The three bulges are the semicircular canals of the vestibular organs. [From Lempert (1948).]

of the malleus are also visible but the stapes is deep in shadow. Running behind the malleus is the chorda tympani nerve which carries the sense of taste. It was mentioned in Chapter 2 that such operations disturb this nerve, which often results in an altered taste sense. In this figure, the rear bony wall of the exterior auditory meatus has been removed, exposing the skin of the meatus which will be used to form a flap to cover the canal after the fenestra has been made. The incus is removed and the head of the malleus is cut off. An opening is then made in the lateral semicircular canal by first thinning the bone of the canal with a dental burr and then cutting out a cap of bone. This method avoids the accidental introduction of bone dust or chips into the canal. This stage of the operation is shown in Fig. 5.52, in which the removed bone cap is also shown. The opening is then covered with a skin flap which consists of a portion of the wall of the

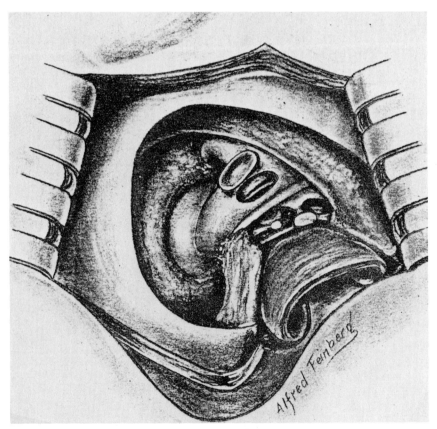

FIG. 5.52 The fenestration operation is the removal of a portion of one of the semicircular canals and covering the opening with a skin flap. [From Lempert (1948).]

meatus together with a part of the drum membrane. Many variations of this operation are possible and fenestra can be made in any of the three semicircular canals. Generally, improvement in hearing is to within 20–30 dB of normal. There are a variety of theories as to why fenestration works, and the interested reader is referred to Wever and Lawrence for a summary of these.

REFERENCES

Ades, H. W. (1959). Central auditory pathway, in "Handbook of Physiology" (J. Field, H. W. Magoun, and V. E. Hall, eds.), Vol. I, Section I, Neutrophysiology, p. 585. American Physiological Society, Washington, D. C.

Bredberg, G. (1977). In "Psychophysics and Physiology of Hearing" (E. F. Evans and J. P. Wilson, eds.). Academic Press, New York.

Brödel, M. (1940). "Year Book of the Eye, Ear, Nose and Throat." Year Book Publishers, Chicago, Illinois.

Brödel, M. (1946). "The Anatomy of the Human Ear." Saunders, Philadelphia, Pennsylvania.

Burke, W., Kotowski, P., and Lichte, H. (1935). Die Hörbarkeit von Laufzeitdifferenzen, *Elektrotechn. Nachr.-Techn.* **12**, 355.

Dallos, P. (1973). "The Auditory Periphery: Biophysics and Physiology." Academic Press, New York.

Davis, H. (1953). Acoustic trauma in the guinea pig, *J. Acoust. Soc. Am.* **25**, 1180.

Davis, H. (1957). Biophysics and physiology at the inner ear, *Physiol. Rev.* **37**, 1.

Davis, H. (1961). Some principles of sensory receptor action, *Physiol. Rev.* **41**, 391.

Davis, H. (1965). A model for transducer action in the cochlea, *Cold Spring Harbor Symp. Quant. Biol.* **30**.

Davis, H., Fernandez, C., and McAuliffe, D. R. (1950). The excitatory process in the cochlea, *Proc. Nat. Acad. Sci. U.S.* **36**, 580.

Davis, H. *et al.* (1953). *J. Acoust. Soc. Am.* **25**, 1180.

Davis, H., Deatherage, B. H., Eldredge, D. H., and Smith, C. A. (1958). Summating potentials of the cochlea, *Am. J. Physiol.* **195**, 251.

de Vries, H. L. (1948). Brownian movement and hearing, *Physica* **14**, 48.

Erulker, S. D., Rose, J. E., and Davies, P. W. (1966). Single unit activity in the auditory cortex of the cat, *Bull. Johns Hopkins Hosp.* **99**, 55.

Gabor, D. (1946). Theory of communication, *J.I.E.E.* **93**, 429.

Gabor, D. (1947). Acoustical quanta and the theory of hearing, *Nature (London)* **159**, 591.

Galambos, R. (1954). Neural mechanisms of audition, *Physiol. Rev.* **34**, 497.

Gray, H. (1974). "Anatomy, Descriptive and Surgical." Running Press, Philadelphia, Pennsylvania.

Guyton, A. C. (1971). "Textbook of Medical Physiology," 4th ed. Saunders, Philadelphia, Pennsylvania.

Helmholtz, H. L. F. (1954). "On the Sensations of Tone" (1885). Dover, New York (reprint).

Kinsler, L. E., and Frey, A. R. (1962). "Fundamentals of Acoustics," 2nd ed. Wiley, New York.

Kopp, C. A., and Green, H. C. (1946). Phonetic principles of visible speech, *J. Acoust. Soc. Am.* **18**, 76.

Kryter, K. D. (1970). "The Effects of Noise on Man." Academic Press, New York.

Lempert, J. (1948). *Proc. R. Soc. Med.* **41**, 617.

Littler, T. S. (1965). "The Physics of the Ear." Pergamon, Oxford.

Morse, P. M. (1948). "Vibration and Sound." McGraw-Hill, New York.

Naftalin, L. (1965). Some new proposals regarding acoustic transmission and conduction, *Cold Spring Harbor Symp. Quant. Biol.* **30**, 169.

Pestalozza, G., and Davis, H. (1956). Electric responses of the guinea pig ear to high audio frequencies, *Am. J. Physiol.* **185**, 595.

Potter, R. K. (1945). Visible patterns of sound, *Science* **102**, 463.

Potter, R. K., Kopp, G. A., and Green, H. C. (1947). "Visible Speech." van Nostrand-Reinhold, Princeton, New Jersey.

Ranke, O. F. (1942). Das Massenverhältnis Zwischen Membran und Flüssigkeit im Innerohr, *Akust. Z.* **7**, 1.

Rosenblith, W. A., and Rosenzweig, M. R. (1952). Latency of neural components in round window response to pure tones, *Fed. Proc.*. **11**, 132.

Shower, E. G., and Biddulph, R. (1931). Differential pitch sensitivity of the ear, *J. Acoust. Soc. Am.* **3**, 274.

Stevens, S. S., and Davis, H. (1938). "Hearing, its Psychology and Physiology." Wiley, New York.

Stewart, G. W. (1931). Problems suggested by an uncertainty principle in acoustics, *J. Acoust. Soc. Am.* **2**, 325.

Tasaki, I. (1954). Nerve impulses in individual fibers of guinea pig, *J. Neurophysiol.* **17**, 97.

Tasaki, I., Davis, H., and Legouix, J. P. (1952). The space-time pattern of the cochlear microphonics (guinea pig) as recorded by differential electrodes, *J. Acoust. Soc. Am.* **24**, 502.

Tasaki, I., Davis, H., and Eldredge, D. H. (1954). Exploration of cochlear potentials in guinea pig with a microelectrode, *J. Acoust. Soc. Am.* **26**, 765.

Towe, A. L. (1965). Audition and the auditory pathway, *in* "Physiology and Biophysics" (T. C. Ruch and H. D. Patton, eds.). Saunders, Philadelphia, Pennsylvania.

von Békésy, G. (1943). Über die Resonanzkurve und die Abklingzeit der Verschiedenen Stellen der Schneckentrendwand, *Akust. Z.* **8**, 66.

von Békésy, G. (1947). The variation of phase along the basilar membrane with sinusoidal vibrations, *J. Acoust. Soc. Am.* **19**, 452.

von Békésy, G. (1949a). The vibration of the cochlear partition in anatomical preparations and in models of the inner ear, *J. Acoust. Soc. Am.* **21**, 233.

von Békésy, G. (1949b). On the resonance curve and the decay period at various points on the cochlear partition, *J. Acoust. Soc. Am.* **21**, 245.

von Békésy, G. (1951a). D. C. Potentials and Energy balance of the cochlear partition. *J. Acoust. Soc. Am.* **23**, 576.

von Békésy, G. (1951b). Microphonics produced by touching the cochlear partition with a vibrating electrode, *J. Acoust. Soc. Am.* **23**, 29.

von Békésy, G. (1952). dc resting potentials inside the cochlear partition, *J. Acoust. Soc. Am.* **24**, 72.

von Békésy, G. (1957). The ear, *Sci. Am.* **197**, Aug. 232.

von Békésy, G. (1960). "Experiments in Hearing." McGraw-Hill, New York.

von Békésy, G. (1962). The gap between the hearing of external and internal sounds, *Symp. Soc. Exp. Biol.* **16**.

Wersäll, J., Flock, A., and Lundquist, P.-G. (1965). Structural basis for directional sensitivity in cochlear and vestibular sensory receptors, *Cold Spring Harbor Symp. Quant. Biol.* **30**, 115.

Wever, E. G., and Lawrence, M. (1954). "Physiological Acoustics." Princeton Univ. Press, Princeton, New Jersey.

Zwislocki, J. J., and Kletsky, E. J. (1979). Tectorial membrane: a possible effect on frequency analysis in the cochlea, *Science* **204**, 639.

CHAPTER 6

Vision

INTRODUCTION

The eye is a sensing device of remarkable complexity, even in primitive forms of life. In vertebrates the visual system begins with a lens for the collection and refraction of light. The lens changes its focal length automatically, depending upon the distance of the object on which the animal focuses. The iris opens and closes to adjust to the intensity of the light, and an internal pigment forms or bleaches in a few minutes for much the same reason.

The formation of the image on the retina must then be translated into an image in the brain. The eye has two end-organs to facilitate this. One system of receptors, called *cones*, has evolved for daylight function when illumination is bright. Cones see detail as well as color. In situations of low illumination, the cones are inoperative and a second type of end-organ, the rods, take over the visual process. These are extremely sensitive and, as we shall see, they can detect but a few quanta of light. With such sensitivity both detail and color are sacrificed.

The spectral range of the eye is less than one octave of frequency, ranging from about 3600 to about 6800 Å (360–680 nm, where

1 nanometer (nm) = 10^{-9}m). Below 360 nm, the ultraviolet region, the energy of the quantum $E = h\nu$, where h is Planck's constant and ν is the frequency, is sufficient to cause electron excitation in organic molecules. Such excitation can obscure the lower energy conformal transformation of a specialized molecule which gives the visual signal. Above 680 nm is the infrared region. Thermal effects in the infrared can cause spurious random excitation of the chemical vision process, and such "noise" can reduce the sensitivity of the eye by lowering the signal-to-noise ratio.

Between the photoreceptors, the rods and cones, and the optic nerve to the brain lie several stages of neurons which analyze the image. All of these neurons are within the retina. With other senses such analysis customarily is a function of the brain. It is therefore sometimes said that the retina of the eye evolved as an extended part of the brain.

In this chapter we will omit the geometrical optics of the eye, which is classical physics and is covered in many texts. [See Bennett and Francis (1962).] We will instead trace the process, as it is now understood, from the receptor to the brain.

EYE ANATOMY

The eye is an approximate spherical ball whose main features for vision are the cornea, iris, lens, and retina. The first three of these focus light on the retina which contains photoreceptors. These, in turn, convert light quanta into electrical impulses, which the optic nerve transmits to the brain, Fig. 6.1.

The cornea is a protective cover for the lens but it also serves to focus light on the retina. The lens in vertebrates adjusts the focusing for near and far vision. By a combination of small muscles around the periphery, the lens is kept in tension, which is a shape with large radius of curvature. When these muscles relax, the lens approaches a more spherical shape, much as a balloon filled with water, Fig. 6.2. The lens, however, is built up in thin layers like an onion. The inner layers are the oldest and, with increasing age, become more separated from the blood supply. Thus, not only is optical accommodation, i.e., change in radius of curvature, decreased with age, but also the inner parts tend to have an amber tint, called yellowing, with age. Because of this, one becomes less sensitive to blue light. It has been suggested that this may be the reason many artists switch from blue as a dominant color in their early works to red as they become older.

The iris is also under involuntary control and can vary the pupil diameter from about 1.5 to 8 mm. Since area is proportional to the square of the diameter, the variation of the pupil area is about a factor of 30. It

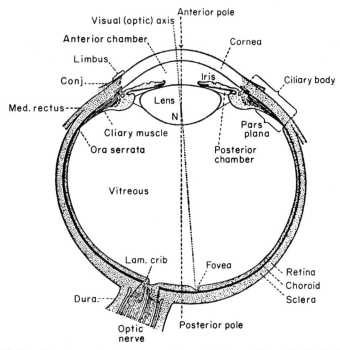

FIG. 6.1 View from above of cross section of right eye. [From Wolff (1954).]

will be seen that vision has a brightness range of 10^5, which is clearly not all due to the pupil area change. Instead, when incoming light is of low intensity, the pupil widens to admit more light. With a wider pupil, the light no longer will fall solely on the most receptive part of the retina, but over a wider region, and in this way the organism trades visual acuity for sensitivity. The pupil also behaves like a camera in changing the f-stop value; the smaller the opening, the greater the depth of field.

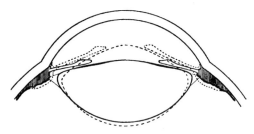

FIG. 6.2 The changes in shape and position of the lens during accommodation. ———, relaxed conditions; ---, accommodated position. [From Starling and Evans, 1962).]

The nerve fibers from the retina all leave the eyeball through the optic disk. There is no photosensitive surface at this point and it therefore constitutes a blind spot. The segment of the retina on the optic axis is called the *fovea* (Latin: little pit), and in this small area of about 1 mm² there is the greatest retinal sensitivity because most of viewing is along the optic axis.

The resolving power of the eye is related to the angle subtended on the retina by an image. Figure 6.3 illustrates the reason for this. If the angle drawn on the retina in this figure is the limit of resolving power, then this angle is given by $\theta = \arctan S/d$, where S is the size of the object and d is the distance. It is obvious that if this ratio is maintained, the eye can distinguish a larger object at a further distance or a smaller one at a shorter distance, provided the object remains outside of the focal length. The usual test is the chart of letters of decreasing size. This is called a Snellen chart, Fig. 6.4 (greatly reduced). With each size letter is a number which indicates the distance the eye must be from it so that the distinguishing part of the letter subtends 1 min of arc on the retina. This part of a letter is, for example, the gap in the letter C which distinguishes it from the letter O. Thus, if a person can only read a line marked 12 m while standing 6 m away, his visual acuity is called $\frac{6}{12}$ or $\frac{1}{2}$. It is now known that visual acuity is a function of the brightness of the object, so most of these charts have been replaced with viewing machines that have controlled brightness.

Visual acuity of the eye is defined as the reciprocal of the resolving power in minutes of arc. Thus, if a test showed a resolving power of 0.5 min of arc, the visual acuity would be 2. The Snellen chart is somewhat crude because of the nonuniformity of the size of the spacings required to distinguish letters and more precise letters are used in viewing machines. As mentioned, there is an effect of illumination. Table 6.1 shows this effect with the luminance expressed in millilamberts (to be defined later).

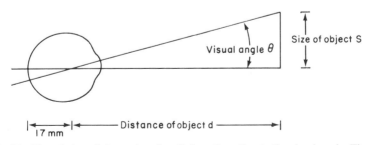

FIG. 6.3   The relation of the angle subtended on the retina to the visual angle. The apex within the eye where the lines cross is about 17 mm in front of the retina and 7 mm behind the foremost point of the cornea.

FIG. 6.4 The Snellen chart reduced in size.

TABLE 6.1[a]

Luminance (mL)	Visual acuity
0.001	0.04
0.01	0.12
0.1	0.40
1.0	1.0

[a] From Starling and Evans (1962).

In the center of the fovea, where visual acuity is highest, the diameter of a single cone is about 1.5 μm. There is a point in any refractive system at which light passing through that point will not be deflected from its path. In the average normal human eye, this point is 17 mm in front of the retina when the eye is focused on a distant object (Fig. 6.3). (When the eye is focused on a near object, the change in refractive power of the lens reduces this distance to 14 mm). Thus, the ratio of the distance of a cone to this distance, 1.5 μm/17 mm, is the tangent of the angle of maximum theoretical visual acuity, and this angle is 20 sec of arc. Since this is $\frac{1}{3}$ min of arc the visual acuity is 3. Under ideal conditions visual acuities of this order are obtained.

THE RETINA

The outer surface of the retina is black from a distribution of pigment cells which contain melanin, the dark pigment of skin. This pigment reduces the scattering of light for the same reason the interior of a camera

FIG. 6.5   Electron micrograph of a fractured portion of the retina showing the rod cells to the left synapsing with other cells. The direction of light from the cornea is from right to left. [Courtesy S. Carlson and D. Bownds, Univ. of Wisconsin (1977).]

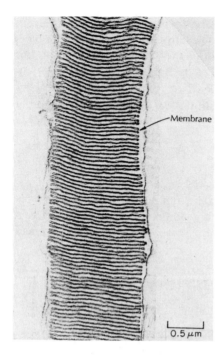

Membrane

0.5 μm

FIG. 6.6   Cross section through the pigment bearing part of a cone from a monkey
retina. Rod structure is similar. [From J. E.
Dowling, *Science* **147**, 57 (1965). Copyright
1965 by the American Association for the
Advancement of Science.]

is painted black. An albino who lacks melanin has great difficulty seeing
contrasts because the scattered light forms a luminous background, much
like trying to see a moving picture show in the daylight.

Inward from this layer, the rod and cone photoreceptors are located.
The rods are seen in Fig. 6.5, which is a photograph of a fractured frog's
retina. The outer part of both the rods and cones have a layered structure.
This is shown in Fig. 6.6. Although the appearance is that of a pile of
disks, the layers are actually sacs or vesicles filled, in the illustrated outer
layers, with a light-sensitive pigment called visual purple or *rhodopsin*. It is
quite possible that this arrangement evolved to increase the area of this
pigment and thereby assure a high probability, or quantum efficiency, of
an incoming photo exciting a photosensitive molecule.

In a human retina there are about 140 million rods and 7 million cones.
The total number of optic nerve fibers is about 1 million, so there is a lot
of convergence of information within the eye before the signal is sent to
the brain. The cones are named for their shape although the shape varies
greatly depending upon their location in the retina.

The distribution of rods and cones in the retina is not uniform, as shown
in the graph of Fig. 6.7. The outer portions of the retina away from the
fovea have only rods, while the center of the fovea has only cones. Figure

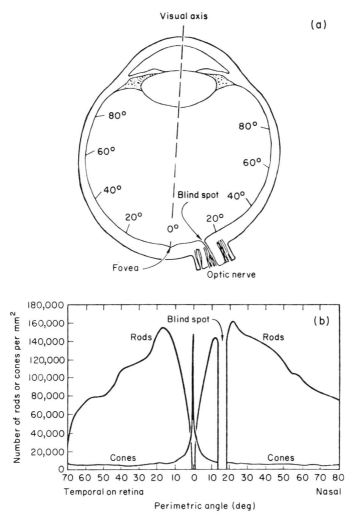

FIG. 6.7   (a) Top view of left eye and (b) the corresponding densities of receptors across the retina. [From Pirenne (1967).]

6.8 illustrates the area of the fovea and optic nerve drawn from a microscope view of the right eye of a monkey. The fovea is within the small dark circle, and the optic disk is the blood-vessel source on the right. The heavy lines are the arteries and veins which pass through the optic disk, as do the fine lines which are the nerves. In the center of the fovea, the cones are so tightly packed that they are smaller than those farther away from the center. An end view of the cones in the human retina at the

160

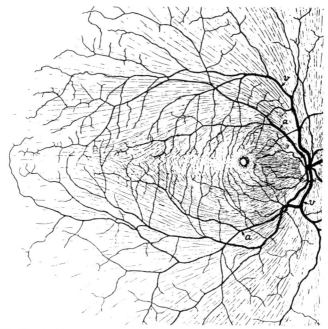

FIG. 6.8 The inside appearance of the retina of the right eye of a monkey as viewed through the cornea. The heavy lines are arteries (*a*) and veins (*v*) which pass through the optic disk on the far right. The fine lines are optic nerve fibers arising from the retinal ganglion cells which gather from all points toward the optic disk where they leave the eyeball and form the optic nerve. The small circle one-quarter of the figure from right is the fovea. Note that neither nerves nor blood vessels interfere with its reception. [Reprinted from S. Polyak, "The Vertebrate Visual System" by permission of The University of Chicago Press. Copyright 1957.]

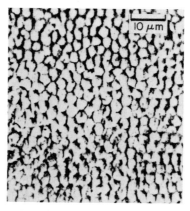

FIG. 6.9 Microphotograph of a portion of the adult human fovea showing the mosaic of cones in the central, rodless area. [Reprinted from S. Polyak, "The Vertebrate Visual System" by permission of The University of Chicago Press. Copyright 1957.]

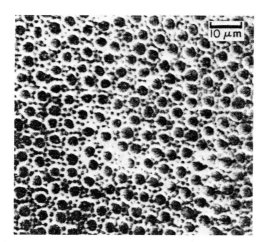

FIG. 6.10 Microphotograph of a portion of monkey retina just outside of the foveal depression. The large circles are cones and the small ones are rods. [Reprinted from S. Polyak. "The Vertebrate Visual System" by permission of The University of Chicago Press. Copyright 1957.]

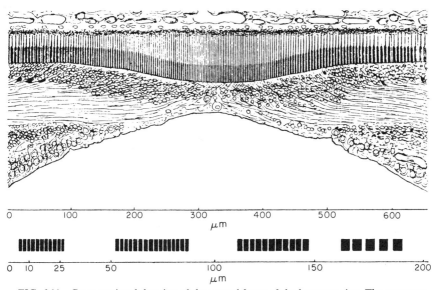

FIG. 6.11 Cross-sectional drawing of the central fovea of the human retina. The receptors are the parallel rectangles in the upper portion. The direction of light is upward. Note that the lateral displacement of all of the cells other than the receptors reduce their interference with the light rays. It is this displacement which creates the depression at the fovea. The lower part of the sketch shows the relative size and number of cones from the center of the fovea (right) to the periphery (left) of the fovea. The accompanying scales indicate actual dimensions in microns. [Reprinted from S. Polyak, "The Retina" by permission of The University of Chicago Press. Copyright 1941.]

center of the fovea is shown in Fig. 6.9. A hexagonal, close-packed mosaic is evident. The density of cones decreases away from the center of the fovea and the density of rods increases. Figure 6.10 shows an enlarged portion of the retina of a monkey 750 μm from the center of the fovea. The large circles are cones and the small block dots surrounding each cone are rods.

It should be noted at this point that the rods and cones are located at the outer portion of the retina, that is, farthest away from the incident light. In order to reach the photosensitive pigment, light must first pass through all of the layers of nerve cells and supporting tissue. This is clearly not an ideal arrangement for sensing light because of absorption. In man and some other species this evolutionary flaw is overcome by the differentiation of the fovea. In this region the inner layers of nerve cells branch away so that the cones are more accessible to the incoming light. It is this branching away of inner nerve cells which causes the pitlike depression of the fovea. This is seen in Fig. 6.11, in which the direction of incoming light is upward. From edge to edge, the fovea is approximately 1500 μm across, subtending an arc of about 5 deg. The rod-free region is about 500–600 μm in diameter.

ILLUMINATION MEASUREMENT STANDARDS

It will be useful at this point to define and quantify the different measurements of illumination. This section may thus be referred to when the terms are encountered in subsequent sections.

It is somewhat unfortunate that serious vision experiments began before the invention of the incandescent light. Candle, or candle power, became the first standard although it is obvious that wick size, wax type, etc., all affect the burning rate and flame size of a candle. As seen below, although the candle remains as a unit, it now has a standard definition.

LUMINOUS INTENSITY

1 candle is the luminous intensity of $\frac{1}{60}$ of 1 cm^2 of surface area of black body radiation at a temperature of 2046°K (the solidification temperature of platinum).

LUMINOUS FLUX

1 lumen (lm) is the luminous flux from a uniform point source of 1 candle through a unit solid angle (steradian), or the flux on a unit surface,

all points of which are at a unit distance from a uniform source of 1 candle. As an example, a 100‑W incandescent bulb emits about 1600 lm.

1 lm of green light $\simeq 4 \times 10^{15}$ photons/sec
1 lm of white light $\simeq 10^{16}$ photons/sec

ILLUMINATION

The illumination of a surface is the luminous flux on the surface. There are three terms in common usage.

(1) 1 footcandle = 1 $\mathrm{lm/ft^2}$
(2) 1 lux (lx) = 1 $\mathrm{lm/m^2}$
(3) 1 phot = 1 $\mathrm{lm/cm^2}$ = 10^4 lx

The phot is used for high intensity illumination and a milliphot, 10^{-3} phot, is sometimes used.

LUMINANCE

Since we rarely look at a light source, but rather on the reflection of the light from objects, the term luminance is used to indicate the brightness of a surface.

1 lambert (L) is the surface brightness of a 100% reflecting and perfectly diffusing surface reflecting light at 1 $\mathrm{lm/cm^2}$.

1 mL = 10^{-3} L.

1 fL is the brightness of a similarly defined surface reflecting light of 1 $\mathrm{lm/ft^2}$.

Some example luminances are given below:

Surface	Millilamberts (mL)
Start-lit sky	2×10^{-4}
Moon-lit sky	7×10^{-3}
Clear sky at noon	$3\text{–}4 \times 10^3$
White paper in sunlight	8×10^3

Retinal luminance

The illumination entering the eye depends upon the size of the opening of the iris, so the luminance must be normalized to the aperture opening of the eye at the time of measurement. 1 troland is the visual stimulation resulting from a luminance of 1 candle/m² when the apparent area of the entrance pupil of the eye is 1 m².

Luminosity

The eye is not equally sensitive to all visible wave lengths (nor are individuals), and a correction factor is necessary when comparing equal or relative brightness of different wavelengths. Quantitatively, luminosity is the ratio of luminous flux to the corresponding energy being radiated expressed in lumens per watt. Most data are given in terms of relative luminosity of different wavelengths of equal luminous flux.

SCOTOPIC AND PHOTOPIC VISION

The cones and rods have different light absorption characteristics. During day vision, called *photopic vision*, the cones are operative while the rods are out of action. Vision near the threshold of illumination is called night vision or *scotopic vision*. At low levels of illumination the cones are not activated. The cones are responsible for color vision, while any illumination in the visible range will activate the rods, but without color sensation. The transition from photopic to scotopic vision is called the *mesopic* range, in which both rods and cones are active. The luminance of this range is about that of full moonlight.

Because of these differences, measurements of visual stimulation of the retina are expressed in terms of scotopic or photopic trolands.

Figure 6.12 shows the wavelength sensitivity of the eye under photopic and scotopic vision. (Note that the relative magnitudes are arbitrary.) As

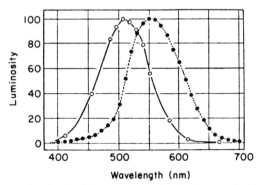

FIG. 6.12   The relative luminosities versus wavelength of scotopic (o) and photopic (●) vision in humans. [From Starling and Evans (1962).]

the sun sets and the brightness decreases, there is a loss of sensitivity to orange and red and a gain in sensitivity to green and blue. This is called the *Purkinje shift* in which, for example, red roses lose their brillance and the color of the green foliage becomes prominent.

ADAPTATION TO NIGHT VISION

If a portion of the retina is excised, it has a purple color until exposed to light for a while, after which it is colorless. The fovea is seen to be yellow colored and does not change upon exposure to light. The purple color is the chemical rhodopsin, or visual purple.

It is known that if a person is in the dark for about 40 min, his night vision acuity increases about a factor of 10,000. This is now known to arise from the formation of rhodopsin in the rods. Thus, rhodopsin forms in the absence of light and transforms, or bleaches, in the presence of light. If one looks at a white cloud on a bright day about one-half of the rhodopsin will be bleached and, since we cannot normally use the rods unless about 90% of the rhodopsin is present, they do not function in light intensities that can activate the cones. The loss of 50% rhodopsin is not expected to reduce the quantum efficiency of excitation of the rods but nevertheless they are not excited at this rhodopsin level. The reason for this is not known.

The bleaching and regeneration rate of rhodopsin have been measured by Campbell and Rushton (1955). They reflected a beam of light from the retina, and the light therefore passes through the rhodopsin twice, once from the incoming beam and then from the reflected beam. The light causes bleaching of the rhodopsin, but the hemoglobin and the melanin of the pigment layer of the retina remain unchanged. Thus, a change in the reflected light is due to visual pigments only and in the peripheral part of the retina, where only rods are present, the changes are due to rhodopsin only. By an ingenious use of a purple wedge in the beam path, an adjustment could be made to maintain a constant amplitude of a photomultiplier output. Figure 6.13 shows some of the results in terms of the purple wedge displacement, the greater the wedge displacement the more rhodopsin present. The curve of solid circles shows the regeneration rate when the rhodopsin is almost fully bleached. This illustrates why it takes so long for the eye to adjust to darkness. The open circles show the bleaching rate for three different intensities of light (1 unit = 20,000 trolands). It is seen that with 1-unit illumination the bleaching rate, while essentially exponential in form, saturates at a level of about 5 wedge units. This saturation occurs because of the chemical equilibrium present between bleaching and regeneration. The 5-unit illumination rate achieves a

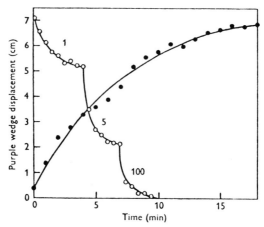

FIG. 6.13   Bleaching and regeneration of rhodopsin in the human eye. ●, regeneration in the dark; o, bleaching under steady retinal illumination of 1, 5, and 100 units of 20, 000 trolands. [From Campbell and Rushton (1955).]

different level of the chemical equilibrium, while the 100-unit illumination effectively overwhelms the regeneration rate and causes almost complete bleaching.

ROD SPECTRAL SENSITIVITY

The eye has different sensitivities to different wavelengths of light. Wald (1945) has measured these for both rods and cones, and the results are shown in Fig. 6.14. At present we are concerned only with rods; the cone data will be discussed later. It is seen, as mentioned earlier, that rods are more sensitive than cones by orders of magnitude except at the red end of the spectrum. The relative sensitivity of the rods has been measured under dark adapted (scotopic) conditions. The maximum sensitivity of the rods occurs at about 500 nm (5000 Å), which is a blue-green color similar to the "green" of a traffic light. The range of vision arises from the molecular nature of the photoreceptor process. Photons of longer wavelength than red light would have insufficient energy to excite the photopigment into a different energy state, while wavelengths shorter than indigo would be able to excite molecules other than the photopigments.

When rhodopsin is extracted from a human retina, its photoabsorption spectrum can be measured independently of its relationship to the retina. The solid curve plotted in Fig. 6.15 is the rhodopsin absorption spectrum. The circles are the spectral sensitivity of a dark-adapted eye as measured

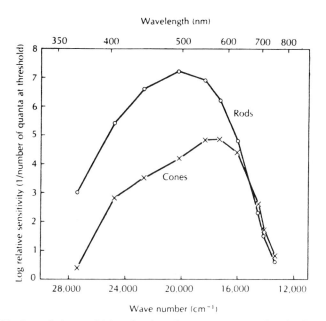

FIG. 6.14   Log relative sensitivity of rods and cones versus wavelength. [From G. Wald, *Science* **101**, 653 (1945).]

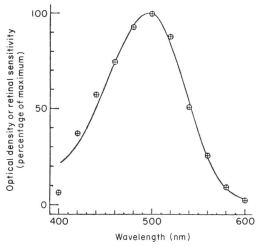

FIG. 6.15   ——, the absorption spectrum of rhodopsin *in vitro*; ⊕, spectral sensitivity of the dark-adapted eye. [Reprinted by permission from F. Crescitelli and H. J. A. Dartnall, *Nature* **172**, 195 (1953). Copyright © 1953 Macmillan Journals Limited.]

by Crawford (1949). This curve is quite similar to that of Fig. 6.14. However, this plot is the percentage of maximum for both the retina and the rhodopsin. What is evident here is that rhodopsin is clearly the photosensitive material of the rods.

TIME AND SPATIAL SUMMATION

The rods sum the light for a time of about 0.1-sec duration before signaling the brain. This was measured by Graham and Margaria (1935) in the following way. The eyes of a subject were dark-adapted for 40 min and one eye was then exposed to flashes of light of an intensity required for observing it 60% of the time, thus just at the threshold of visibility. The test spot was 2 min of arc diameter on the retina and 15 deg to the side of the optic axis, therefore on rods alone. The only variable was the duration of the flash. The results are shown in Fig. 6.16. It is seen that as long as the flash is shorter than about 0.1 sec, its duration has no effect and the flash is seen. If the duration is greater than 0.1 sec, then a greater intensity of light is required for visibility. All of the quanta required for threshold visibility can be delivered over a 0.1-sec interval or in a 10^{-6}-sec interval without causing a noticeable difference in the results. Furthermore, one-half of the required quanta can be delivered in 10^{-6} sec and the other half in a similar flash at any later time up to 0.1 sec without a change. This

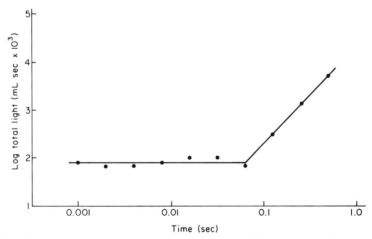

FIG. 6.16 Log of the product of intensity and time (It) versus time (t) for a 2-min diameter arc on the human retina. Measurements made at threshold on a dark-adapted eye. [Data from Graham and Margaria (1935).]

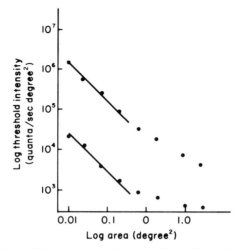

FIG. 6.17 Log threshold intensity versus log area of retinal stimulation for dark-adapted eye with no background illumination. Upper curve, 8.5×10^{-3}-sec duration; lower curve, 930×10^{-3}-sec duration. [Data from Barlow (1958).]

means that the effect of a single quantum absorption must last for about 0.1 sec so that its effects can be added to that of any other quantum absorbed within that time. Note that up to about 0.1 sec the curve is flat, or It = const, while above 0.1 sec the slope It/t yields I = const.

The mosaic arrangement of receptors in the retina has an effect on their connections to the nerve signals, although the details are not known. What is known is that, up to a certain size, the retina cannot distinguish the size of a spot of light. This measurement was performed by Barlow (1958), in which the variable was the retinal diameter of a light at threshold intensity on a dark-adapted eye. The flash durations were 8.5×10^{-3} and 930×10^{-3} sec, which are within the temporal summation measurements above. Figure 6.17 shows the results. In this figure the threshold intensity per unit area is plotted against the illuminated area of the retina. The upper curve is the short illumination, 8.5×10^{-3} sec, and it is seen by the straight line that there is no effect of area up to about 0.4 deg^2 (44 min of arc diameter). Barlow also showed in the same experiments that as the background illumination is increased, the area for which there is no change is considerably reduced. The decrease in slope as the area increases shows that the threshold is not as sensitive, that is, the effectiveness per unit area is reduced. This is because adjacent groups of rods do not enhance the visual process but actually inhibit it. This effect of inhibition will be discussed later in this chapter.

THE STEREOCHEMISTRY OF RHODOPSIN

In 1933 Wald isolated vitamin A from the retina and commenced a four-decade long study of its role in vision. Night blindness is an early symptom of vitamin A deficiency, although the time of onset of night blindness varies with individuals. This is because vitamin A is stored in the liver and the stored quantity can supply visual needs for some time. The results of careful studies of this on rats are shown in Fig. 6.18. In this figure various biophysical measurements are shown on the ordinate and time in weeks on the abscissa. At the beginning, a vitamin A deficient diet was commenced. It is seen that the vitamin A content of the liver, which has continued to supply the retina through the blood, falls to zero in three weeks. Shortly afterward, the supply in the blood falls to zero and, as that decreases, so does the rhodopsin content of the eye. When that has fallen 40%, the opsin concentration of the retina begins to decrease with associated histological deterioration of the retina.

When large doses of vitamin A are then administered, recovery can take place rather dramatically. This is seen in Fig. 6.19. The three curves are for three different animals and the ordinate is the logarithm of the threshold illumination required for night vision. The threshold level drops, in the most striking case, by a factor of 1000 in two days, returning to normal.

Actually, in the retina vitamin A is in its oxidized form, called *retinal*, and this is found in every species from clams to man. There are two forms of vitamin A, vitamins A_1 and A_2, which differ mainly in that A_2 has an additional double bond in the ring, called the β-ionone ring, shown in

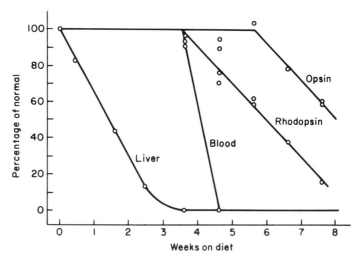

FIG. 6.18 Biochemical changes in a group of white rats on a vitamin A deficient diet. [From Dowling and Wald (1958).]

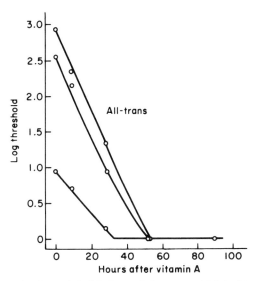

FIG. 6.19 Recoveries from night blindness of three rats which had been on a vitamin A deficient diet measured at various times after intraperitoneal injection of large doses of all-trans vitamin A_1. [From Dowling and Wald (1956).]

Figs. 6.20a and 6.20c, and therefore there are also two forms of retinal, shown in Figs. 6.20b and 6.20d.

A linear molecule of this type can have other shapes or configurations. A portion of the chain can be rotated so that while the same angle between the single and double bonds is maintained, it may occur on the other side of the carbon. The linear configuration of Fig. 6.20 is called all-trans. A cis

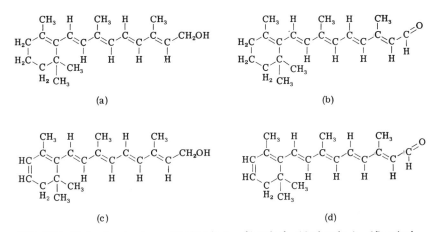

FIG. 6.20 Molecular structures: (a) vitamin A_1, (b) retinal$_1$, (c) vitamin A_2, (d) retinal$_2$.

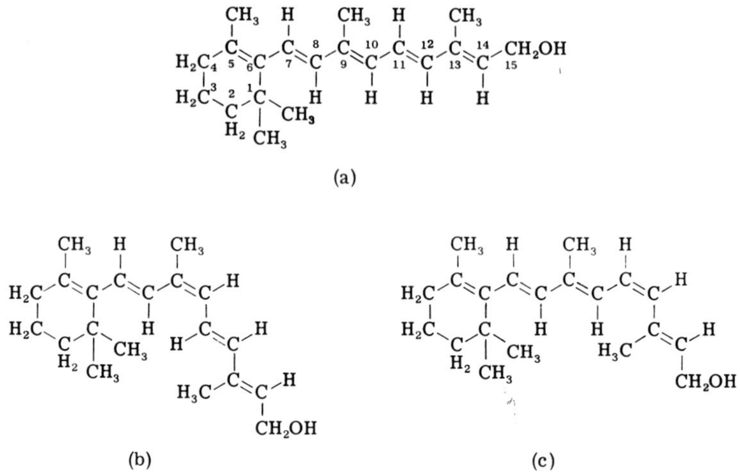

$$
\begin{array}{l}
\text{structure (a) vitamin A}_1 \text{ all-trans}
\end{array}
$$

(a)

(b)   (c)

FIG. 6.21   Molecular structures: (a) vitamin A_1 all-trans (b) 9-cis, (c) 11-cis.

configuration is preceded by a number, the number of the carbon atom about which the cis transformation takes place. The numbering of the carbon atoms in vitamin A_1 is shown in Fig. 6.21a. Two of the several possible isomers are shown in Figs. 6.21b and 6.21c. The transformation from trans to cis requires a rotation about a bond. If other atoms are in the way, it cannot be done without extra energy. Furthermore, in the cis configuration, if some atoms are too close together the molecule cannot exist in a planar form. Pauling (1949) has shown that one can expect to find the cis configuration only in the double bonds adjacent to the methyl (CH_3) groups, 9 and 13 in Fig. 6.21a. At 7 or 11, the molecule would encounter overcrowding (steric hindrance) in going from the trans to the cis configuration. For example, a cis linkage at 7 would bring the methyl groups on carbons 1 and 9 into an overcrowded position. Similarly, in a cis linkage at carbon 11, the hydrogen on carbon 10 would overlap the methyl group on carbon 13. This is illustrated in Fig. 6.22, in which the small circle represents the space occupied by the hydrogen atom and the large circle that occupied by the CH_3 group. Because of this molecular over-crowding, or steric hindrance, the 11-cis configuration cannot exist in a

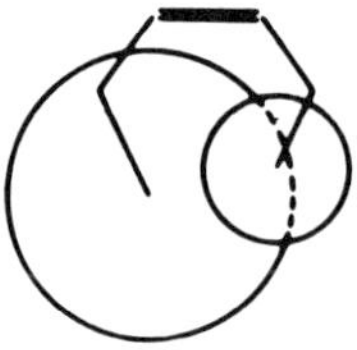

FIG. 6.22   Steric hindrance of an H (small circle) with a CH_3 (large circle). [From Pauling (1949).]

planar form but must be twisted out of the plane. Pauling's original estimate (1949) was that the energy of distortion was so large that the configuration would not be present in an equilibrium mixture at room temperature. Later experiments by Hubbard (1966) showed that the ground state free energy of 11-cis was only 1.5 Kcal/mole greater than the all-trans isomer.

It is the 11-cis isomer which nature has selected to use for visual systems. Image resolving eyes occur in only three phyla: mollusks, arthropods, and vertebrates. The eyes of these three are quite different anatomically, and presumably each phylum had its own evolutionary path. However, all three utilize vitamin A and the 11-cis to all-trans transformation. As will be seen, the interaction of 11-cis retinal with the protein opsin makes it very useful.

Opsin is a protein molecule with a molecular weight of about 38,000. Its structure has not yet been completely determined. There are two types of opsins in vertebrates, one in rods and the other in cones. The two types of retinals join with the two types of opsins to yield four pigments. Table 6.2 shows the combinations and their names as well as the wavelength of maximum absorption. We will consider only rhodopsin, for it has been recently shown that the different chromophores are formed from this (p. 248).

TABLE 6.2[a]

				λ_{max}(nm)
	+ rod opsin	→	rhodopsin	500
retinal$_1$	+ cone opsin	→	lodopsin	562
	+ rod opsin	→	porphyropsin	522
retinal$_2$	+ cone opsin	→	cyanopsim	620

[a] From Wald (1961).

The small, light-sensitive molecule retinal in the 11-cis configuration is bound to the large opsin molecule in a protonated Schiff-base linkage. This combination of molecules is called a *chromophore*. In this linkage an H_2N on the opsin molecule exchanges its N for the O on the retinal molecule with a molecule of water being released. In addition, a proton is added to the linkage at the nitrogen position. Schematically the ends of the retinal and opsin molecules have the following reaction.

$$R\text{-}C \overset{\displaystyle /\!\!/ O}{\underset{\displaystyle \backslash\, H}{}} + H_2N\text{-}opsin \overset{H^+}{\longrightarrow} R\text{-}C \overset{H}{\underset{}{|}} = \overset{H^+}{\underset{}{N}} \text{-}opsin + H_2O$$

$$11\text{-}cis\ retinal + \frac{lysine\ side}{chain\ of\ opsin} \rightarrow protonated\ Schiff\ base + water$$

In addition to this linkage, Matsumoto and Yoshizawa (1975) have shown that the β-ionone ring also binds to a site in the opsin at a faster rate than the Schiff-base linkage.

The opsin plays no role in the light absorption, only the retinal is affected. However, the opsin does have some effect on the absorption spectra. The Schiff-base linkage alone has a maximum absorption wavelength of 360 nm, and protonated it has $\lambda_{max} = 400$ nm. This is not large enough to explain the 500-nm maximum exhibited by rhodopsin. It is known, however, that if a protonated Schiff-base linkage is in a local environment a highly polarizable shift to the red of the absorption spectrum is quite possible. In other species the opsin environment probably provides a somewhat different polarizability which accounts for the differences in their absorption spectra. See Honig et al. (1975) for a review of the absorption characteristics. See also p. 248.

When a quantum of light is absorbed by a rhodopsin molecule, it isomerizes the retinal from 11-cis to all-trans form. This changes both the shape and the dimension of the molecule so that it no longer fits into the binding site of the opsin molecule. Actually, there are several stages in this transformation, each of which has a reaction time and a temperature at which it freezes so that the stages can be studied in detail. These will be discussed in a later section. For now, Fig. 6.23 symbolizes the possible steps in which rhodopsin, upon absorbing a photon, isomerizes to *lumirhodopsin*, whose activity is sufficiently slowed below $-45°C$ to permit its study. The next stage is *metarhodopsin* which can be studied below $-20°C$. With the addition of H_2O, it hydrolizes to retinal (retinene in the figure). Later and more detailed studies of this process are discussed on p.

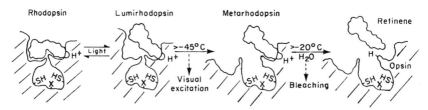

FIG. 6.23 The action of light on rhodopsin. The absorption of light by rhodopsin isomerizes its 11-cis chromophore to the all-trans configuration, yielding as first product the all-trans chromoprotein lumirhodopsin. This labilizes the protein, opsin, which rearranges to a new configuration, yielding a second all-trans chromoprotein, metarhodopsin. This second process exposes reactive groups on opsin; two—SH groups, and one proton-binding group, symbolized in the figure with a negative charge—and may be responsible for triggering visual excitation. Vertebrate metarhodopsins are unstable, and above about $-20°C$ hydrolyze to opsin and all-trans retinal, the process that corresponds to bleaching. [From Hubbard and Kropf (1959).]

207. As retinal, it is yellowish in color and does not absorb light. It must undergo a chemical reaction to be reconverted into rhodopsin. This is the process of the bleaching of rhodopsin discussed in an earlier section.

ABSOLUTE SENSITIVITY OF THE RETINA

The electrical impulse from the retina to the optic nerve is initiated by external energy activating a photoreceptor. This energy is that contained in a quantum of light through the relation $E = h\nu$, where ν is the frequency of the wavelength of the light and h is Planck's constant, 6.62×10^{-34} J/sec (6.62×10^{-27} erg/sec). Very distant stars are visible to the unaided eye so it is known that only a small number of quanta are required for stimulus. How small a number is required? This measurement was made with great precision by Hecht, et al. (1942), who used two techniques which gave essentially the same answer, although sensitivity varies slightly for different observers.

Hecht et al. were aware of a number of factors which are involved in measuring the maximum sensitivity of the eye. (1) The rods are more sensitive than the cones, but they require almost one full hour in darkness to reach their maximum sensitivity. (2) The maximum sensitivity is not associated with direct vision but rather with peripheral vision, and this is at 20 deg from optic axis of the eye (see Fig. 6.7). (3) The size of the spot focus on the retina must be small (Fig. 6.17). (4) The exposure time must be short because of the fatigue which occurs in both the chemical and electrical system of the eye (Fig. 6.16). (5) The dark-adapted eye is known to be most sensitive to light with a wavelength of 510 nm (Fig. 6.12).

In order to perform the experiments with these conditions the observer, with his eyes dark-adapted, placed his head in a fixed poisition by clenching his teeth in a mold of the bite of his upper jaw. He would stare through a hole with one eye at a faint red light. This color does not change the dark adaption of his retina. Another light of 510 nm was placed at an angle of 20 deg from the axial line of his eye and the red light. The observer flashed this light himself when he felt ready. The intensity of the light was varied by the experimenter. Thus, although the observer controlled the flash, he did not know its intensity and would record only if he saw it or not. The duration was set at 0.001 sec and the intensity was varied by neutral density filters.

The energy reaching the eye was determined by placing a sensitive thermopile at the position of the pupil and correcting the result for the optics of the eye. Thus, the results were based on whether or not a certain amount of energy delivered to the eye in 0.001 sec was detectable or not.

This energy is $E = nh\nu$, where n is the number of quanta and the frequency ν is determined from $c = \lambda\nu$, where c is the velocity of light and λ the wavelength.

Each of a series of intensities was presented many times and the frequency of seeing the flash was determined for each. From the resulting plot of frequency of observation against intensity, the threshold of detectability was chosen as the amount of light that could be seen 60% of the time. Several observers were used over a period of a year and a half, with several determinations for each. For all of these observers the minimum energy at the cornea necessary for detectability ranged between 2.1 and 5.7×10^{-17}J (2.1 and 5.7×10^{-10} ergs). Since $h\nu$ for 510-nm wavelength is 3.89×10^{-19}J/quantum, this range of energies corresponds to 54–146 quanta.

These values, however, do not represent the amount of energy which arrives at the retina. There are three corrections which must be applied. (1) Reflection from the cornea is about 4%. (2) Absorption by the media between the cornea and the retina amounts to 50% of the energy. (3) The retina does not absorb all of the light that strikes it. By separate measurements of the rhodopsin absorption of a frog's eye, these investigators concluded that 20% absorption is the upper limit. These corrections must be applied to the detectable energy arriving at the pupil in order to determine the number of quanta which can be detectable at the retina. Therefore, the 54–149 range of quanta arriving at the pupil must be multipled by the fraction $f = 0.94 \times 0.5 \times 0.2 = 0.094$ and the resulting range of 5 –14 quanta represent the detectable range for different individuals and the same individuals during different experiments.

It is important at this point to consider the angle that the diameter of the blue-green light subtended on the retina in this experiment. As will be discussed later in this chapter, the rods of the retina are grouped into "receptive fields" whose angular size, or viewing angle, is roughly 10 min of arc. If the incident light on a receptive field subtends an arc less than this, the eye's response is the same as if it covered the entire receptive field. If the angle is greater, then the overlapping of the light into other receptive fields can inhibit the response of the illuminated receptive field. Knowing this, Hecht et al. kept the angle of the light on the receptive field within 10 min.

At the illuminated region of the retina, 20 deg from the optic axis, the rod density is highest and the 10-min circular vision field contains about 500 rods. If 5–14 quanta are absorbed by 500 rods, it is statistically improbable that any rod will absorb more than 1 quantum. This means that 5–14 quanta probably change one molecule in the same number of rods in a group of 500 to give a visual effect. This is a very small number of chemical events, but this extreme smallness lends itself to be tested by an entirely different method.

The arrangement of the flash in the preceding experiment involved an incandescent filament with a rotating shutter. The calibration of the thermopile was done with the shutter open and the filters removed, and the energy of each flash was calculated from the shutter time and filter density. Such a calculation gives the average number of quanta which arrive at the pupil. Each flash, however, will not always contain this average number but will have a statistical variation about the average.

The probability of discrete events about an average is given by the Poisson probability distribution (Vol. I, Eq. (C.8))

$$P_r = m^r e^{-m} / r! \qquad (6.1)$$

where P_r is the probability that r events occur when m is the average number of events. Tables of values of P_r for various values of r and m are available. Figure 6.24 shows a plot of P_r for various values of r (ordinate) against log of the average number m (abscissa) in a given flash.

There are two significant features of the curves of Fig. 6.24. One is that the shape of all of the P_r distributions are fixed and vary differently for each value of r. The second is that the variation of P_r is expressed in terms of the logarithm of the average number of quanta in a flash. Therefore, the absolute value of the number of quanta which reach the retina per flash need not be known, but only the relative number. That is, if the absorption fraction of the rhodopsin mentioned earlier is incorrect by a constant multiplicative factor, it does not matter since it is the shape of the curve that is to be matched by a displacement along the abscissa. The experiment then is to do many flashes of a given average energy content, and the frequency of seeing it will depend on the probability that it yields r or

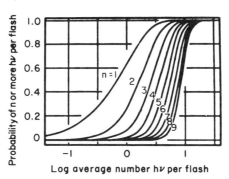

FIG. 6.24 Possion probability distributions. For any average number of quanta ($h\nu$) per flash, the ordinates give the probabilities that the flash will deliver to the retina n or more quanta, depending on the value assumed for n. [From Hecht et al. (1942).]

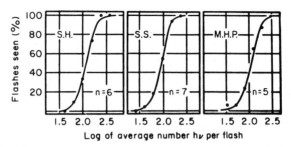

FIG. 6.25   Relation between the average energy content of a flash of light (in number of *hv*) and the frequency with which it is seen by three observers. Each point represents 50 flashes, except for S. H. where the number is 35. The curves are the Poisson distributions of Fig. 6.24 for *n* values of 5, 6, and 7. [From Hecht *et al.* (1942).]

more quanta per flash. This same experiment is repeated for flashes of different energy content (different average quanta content m). In this way, a distribution is obtained whose shape, when plotted against the logarithm of the average quanta content m, should match one of the probability distributions of Fig. 6.24 and thereby obtain uniquely the value of r.

The observer is placed in the same position as before and he triggers the flash when he is ready. The experimenter varies the energy content of the flash at random and the observer reports whether or not he has seen it. Three observers were used in the experiment with the numbers of flashes from 35 to 50. Observers S. H. and S. S. performed the experiment twice. The results are shown in Table 6.3. These percentages are plotted against the logarithm of the average number of quanta per flash in Fig. 6.25. The shapes of the curves are then compared to those of Fig. 6.24. It is seen by this comparison that the three observers are able to see flashes which correspond to numbers of quanta between 5 and 8. These results are in excellent agreement with the experiment described earlier which led to values in the range of 5–14 quanta.

Impaired vision can result from two separate phenomena, based on the above considerations. If the cornea is pigmented, it will absorb some of the incident light so a higher intensity is required to deliver the threshold amount of quanta to the retina. In Table 6.3, it should be noted that S. H. was the oldest of the observers, and it is seen in the data that he required a higher incident energy to record each percentage. However, his retina was of comparable sensitivity to the quanta which did reach it. Damage to the retina, however, can reduce the number of sensitive rods or reduce their quantum efficiency for the photochemical process. In this situation some of the quanta will not initiate the impulse, having fallen on an "infertile field." More quanta are therefore required to have the five or more successful "hits" in a receptor field, which will require a higher intensity light source as a stimulus.

TABLE 6.3

Energy and Frequency of Seeing[a, b]

S. H.		S. H.		S. S.		S. S.		M H. P.	
No. of quanta	Fre-quency per cent	No. of quanta	Fre-quency per cent	No. of quanta	Fre-quency per cent	No. of quanta	Fre-quency per cent	No. of quanta	Fre-quency per cent
46.9	0.0	37.1	0.0	24.1	0.0	23.5	0.0	37.6	6.0
73.1	9.4	58.5	7.5	37.6	4.0	37.1	0.0	58.6	6.0
113.8	33.5	92.9	40.0	58.6	18.0	58.5	12.0	91.0	24.0
177.4	73.5	148.6	80.0	91.0	54.0	92.9	44.0	141.9	66.0
276.1	100.0	239.3	95.7	141.9	94.0	148.6	94.0	221.3	88.0
421.7	100.0	386.4	100.0	221.3	100.0	239.3	100.0	342.8	100.0

[a] Relation between the average number of quanta per flash at the cornea and the frequency with which the flash is seen. Each frequency represents 50 flashes, except for S. H., for whom there were 35 and 40 for the first and second series, respectively.

[b] From Hecht *et al.* (1942).

SENSITIVITY ABOVE BACKGROUND

Obvious questions at this point are the following: Why does a receptive field of the retina require 5–14 photons for detection? Will the eye evolve further in future eons so that it can detect a single photon? The answer is that the eye is fully developed in an evolutionary sense. It requires 5–14 photons in the signal to overcome the background noise. The ideas in this concept of signal-to-noise ratio in vision were first developed by Rose (1942, 1948) and deVries (1943). The experiment which demonstrated the correctness of these ideas was performed by Barlow (1957). This experiment will now be described, and the discussion will follow to some degree the pedagogical summary given by Villars and Benedek (1974).

The stars are visible at night but not during daylight. This is because their incremental light is too faint to see above the background light of the sky when the sun is up. A similar phenomena is noticed when trying to observe photographic slide projection when the room is not adequately darkened. The eye requires a certain incremental intensity ΔI over a background intensity I in order to detect the information. The ratio $\Delta I/I$ is known as Weber's law and will be discussed further in Chapter 7.

Any chemical process, such as the rhodopsin transformation in the retina, requires energy. Not only can this energy be supplied by a photon, it can come from heat. The thermal energy of molecules is distributed among them statistically, and, although at body temperatures very few of the rhodopsin molecules have sufficient energy for a spontaneous excitation, a few of them do. Thus, there is a background "noise" level even in

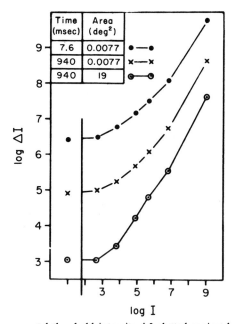

FIG. 6.26   Log incremental threshold intensity ΔI plotted against log background intensity I. ●, for short duration small area; ×, for long duration small area; and ○, for long duration large area, test stimuli. Intensities in quanta (507 nm)/sec deg². [From Barlow (1957).]

the dark. Barlow's (1957) experiment determined that this background noise is equivalent to about 4 quanta per receptive field for each characteristic exposure time (or retina integration time discussed above) of 0.1 sec. Therefore, more than 4 quanta are necessary to see a signal ΔI above this thermal noise background. In addition to this measurement, Barlow determined the $\Delta I / I$ ratio at a variety of background intensities I, the results of which clearly demonstrate the statistical nature of vision.

The experimental arrangement for the subject was similar to that used by Hecht *et al.*, described in the preceding section. The subject's head was in a fixed position, the eyes were dark adapted, a red spot was used for alignment, the measured light was off the axis (in this case by 6 deg) with a small field of about 0.1 deg of arc diameter, and the flash had a duration of a few milliseconds. Two concentric, overlapping circles of light of 507 nm were flashed on the screen. The outer one had an angle of 13 deg. The larger circle was the background intensity I, and the subject varied the intensity of the inner one ΔI until he could see it about 80% of the time. Typical results are shown in the upper curve (solid circles) of Fig. 6.26. Another experiment was performed in which the exposure time was almost

1 sec, and the angle of the inner ΔI circle was about 4.5 deg. These results are shown in the curve with open circles. Its significance will be discussed later.

We have seen earlier, in the discussion of the experiments of Hecht *et al.*, that only a fraction $f = 0.094$ of the light which strikes the cornea can activate molecules in the retina. This number of quanta which excites the retina is called Δn. The maximum retina integration time τ, Fig. 6.16, is 0.1 sec. I is the background intensity in quanta/sec deg^2 entering the receptive field. The area of the receptive field of the retina is Ω deg^2 (Barlow used α in his paper).

With these symbols, we may express the average number of photons $\bar{n}_I$ from the background illumination intensity I which strike an angular area Ω in retina integration time τ as

$$\bar{n}_I = I\Omega\tau f \qquad (6.2)$$

The fraction f is required because I is the intensity which strikes the cornea. The thermal excitation of rhodopsin molecules is equivalent to a steady background intensity x. This can be written as $\bar{n}_x$, the average background intensity

$$\bar{n}_x = x\Omega\tau f \qquad (6.3)$$

to have the same form as Eq. (6.2). The number of excitations per unit retinal integration time τ is the sum of these two terms or

$$\bar{n} = \bar{n}_I + \bar{n}_x = (I + x)\Omega\tau f \qquad (6.4)$$

The flash of light of the small circle which contributes ΔI intensity above this background contributes ΔN photons at the cornea or, the number of retinal excitations

$$\Delta n = f\Delta N \qquad (6.5)$$

Therefore, the sensitivity of the eye must distinguish the presence of Δn photons in the presence of $\bar{n}$ background photons. This is called the signal-to-noise ratio, and we will denote this for threshold detection as K.

For random fluctuations, the mean value is given by the average of the square of the fluctation amplitude, and noise $\mathfrak{N}$ is the root-mean-square fluctuation (See the Appendix). Using these terms in our present case, Δn is the fluctuation amplitude about the mean $\bar{n}$. Therefore,

$$\bar{n} = \overline{(\Delta n)^2} \qquad (6.6)$$

and

$$\sqrt{\bar{n}} = \left(\overline{(\Delta n)^2} \right)^{1/2} \qquad (6.7)$$

Noise $\mathfrak{N}$ is defined in the random fluctuation case as the root-mean-square fluctuation or, from Eq. (6.4),

$$\mathfrak{N} = \sqrt{\bar{n}} = \left[(I + x)\Omega\tau f \right]^{1/2} \qquad (6.8)$$

The signal S which must be detected above this background is

$$S = \Delta n = f\Delta N \qquad (6.9)$$

and the threshold signal-to-noise ratio is

$$K = S/\mathfrak{N} = f\Delta N / \left[(I + x)\Omega\tau f \right]^{1/2} \qquad (6.10)$$

Rearrange this equation to obtain an expression for ΔN,

$$\Delta N = K(\tau\Omega/f)^{1/2}(I + x)^{1/2} \qquad (6.11)$$

On the basis of Eq. (6.11), Barlow plotted data from a short exposure time, small angle experiment, similar to the upper curve of Fig. 6.26 and obtained the circles in Fig. 6.27. Although both x and K are unknown in Eq. (6.11), the shape of the curve can be shifted to match the data. Thus, the solid curve of Fig. 6.27 is seen to match the data, and from this x and K can be evaluated.

The multiplicative constant

$$C = K(\tau\Omega/f)^{1/2} \qquad (6.12)$$

of Eq. (6.11), from Fig. 6.27, is 3.3 deg (sec)$^{1/2}$. We know from the previous discussion that $\tau = 0.1$ sec and $f \cong 0.1$. The effective size of a receptive field of the retina is not a fixed quantity, however, but differs in various

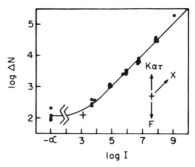

FIG. 6.27 Log increment threshold quantity of light (ΔN) plotted against log background intensity I. Four determinations on one subject. Stimulus short duration (8.6 msec), small area (0.011 deg^2). Thresholds in number of quanta of 507 mμ at the cornea; background intensity in quanta/sec deg^2. Solid line is theoretical curve for $\Delta N = K(\alpha\tau/F)^{1/2} (I + X)^{1/2}$ for $K(\alpha\tau/F)^{1/2} = 3.3$, $X = 1260$. + shows point of intersection of horizontal and slanting asymptotes of theoretical line, and in the bottom right corner arrows show how this point, and with it the whole theoretical curve, is displaced by increases of the different parameters. [From Barlow (1957).]

regions of the retina. Barlow (1958) investigated the size and concluded that the upper limit for the size of the receptive field is about $\Omega = 0.4$ (deg)^2. Inserting these numbers in Eq. (6.12) with the experimental value of C yields a value for the threshold signal-to-noise ratio of $K = 5$. How does this compare with generally accepted signal-to-noise ratios? Clearly, the ratio must be greater than 1, but how much greater depends to some extent the observer and on what fraction of correct discriminations is taken as the threshold value.

Before continuing the discussion of K, let us follow Barlow's evaluation of x when $I = 0$, the "dark light" or intrinsic thermal stimulation of the retina. The theoretical curve of Fig. 6.27 matches the data when $x = 1260$. From the above formulation, this must be 1260 quanta/sec (deg)^2, and substituting this value into Eq. (6.3),

$$\bar{n}_x = x\Omega\tau f$$

$$= \left[1260 \frac{\text{quanta}}{\text{sec deg}^2} \right] \times 0.4 \text{ deg}^2 \times (0.1)\text{sec} \times (0.1)$$

$$\bar{n}_x = 5 \text{ quanta}$$

This is the number of quanta to which the thermal background noise is equivalent. However, this is not a fixed number; it too fluctuates, and, as in Eqs. (6.6) and (6.7), if Δn_x is the fluctuation amplitude

$$\sqrt{\bar{n}_x} = \sqrt{(\Delta n_x)^2} \qquad (6.13)$$

which is the dark current noise $\mathfrak{N}_x$. Therefore

$$\mathfrak{N}_x = \sqrt{\bar{n}_x} = \sqrt{5} = 2.2$$

At the threshold of vision, the magnitude of signal S must be greater than the noise $\mathfrak{N}$ by some factor K

$$S/\mathfrak{N}_x > K \qquad (6.14)$$

With K evaluated at 5 and $\mathfrak{N}_x$ as $\sqrt{5}$, S is greater than 2.2. Barlow reported, however, that with different observers S can have as much as twice the value of that obtained from the observer's data of Fig. 6.27. A multiplicative factor of 2 times x in Eq. (6.10) increases S by $\sqrt{2}$, and from Eq. (6.14) $x = 3$. Thus, a minimum of 3 quanta is required just to overcome thermal noise.

The experiment of Hecht et al. described above found the threshold of sensitivity of a dark-adapted eye to be between 5 and 8 quanta, a value not much greater than the 3 quanta required to exceed thermal background determined by a completely different experiment. It is therefore clear from random fluctuation theory why the eye cannot detect a single quantum, and that evolution has brought the receptor of the eye to its maximum possible sensitivity.

Even the small discrepancy between the 3 quanta of the Barlow experiment and the 5 to 8 of the Hecht *et al.* experiment can be rationalized. Cataracts in the lens of the eye cause a scattering of light, some of which strikes the fovea. In this way, the background intensity I is higher than normal, and a correspondingly higher ΔI is required to "see" or discriminate. This situation is called "glare sensitivity" and makes it difficult, for example, to drive at night. One cannot distinguish the edge of a highway in the glare of headlights of an oncoming car. The normal eye lens probably does not have perfect transmission, and some scattering from pigments is probably always present. Thus, somewhat more than 3 quanta are required for threshold vision, and a correspondingly higher signal-to-noise ratio K is required for a certain confidence level.

SIGNAL-TO-NOISE RATIO IN VISION

The development of the television picture tube led to careful consideration of human vision. Basic studies were made concerning the precise number of separate pixels required for the eye to perceive a clear picture instead of a grainy one. The rate of electrons striking the screen phosphor with respect to its radiative decay rate and the number of excitations of the phosphors required to achieve different shades of gray had to be studied. Since the electrons were countable, as were the individual phosphor excitations, the absolute number of photons could be calculated. Human viewers could then evaluate the required quantum levels, and a realistic signal-to-noise ratio could be obtained. The history of these studies and a summary of the findings is given in the book by Rose (1973), one of the pioneers in the field. We will follow some of his analysis.

Imagine a blackboard divided in a grid pattern of N squares. A single grid unit has a fractional area $1/N$ of the entire grid. If we wish to portray a single black square, then all of the other $N - 1$ squares must be painted white. Suppose we wish instead to portray a single gray spot with 99% reflectivity of the white spots. We must not only locate it as before, but we must portray its shade of grayness. First, in order to introduce the shade, each grid element must be subdivided into 100 dots so that the gray element can have 99 dots. As before, the grid element position must also be specified since each of the N grid elements now contain 100 dots. The required number is obviously $100\,N - 1$. Thus, each element of the grid will have 100 dots and the gray one 99. This shows that to indicate 100 different shades of gray, the number of photons must be increased by a factor of 100.

In any beam of photons one may count the numbers in separate, equal time units Δt, with n_1 in Δt_1, n_2 in Δt_2, n_i in Δt_i, etc. Even though the Δt's are equal, in a probability process the n's will not be equal because of a certain randomness. If we average the number of photons per time element and call this $\bar{n}$ then, as given in Eq. (A.1), $\bar{n}$ is also equal to the average value of $(n_i - \bar{n})^2$, or

$$\bar{n} = \overline{(n_i - \bar{n})^2} \tag{6.15}$$

This term is called the mean-squared deviation from the mean and its square root $[(n_i - \bar{n})^2]^{1/2}$ is called the root-mean-squared deviation, or rms deviation. The terminology in use defines a *signal* S as the average number of photons striking a test element and *noise* $\mathfrak{N}$ as the rms deviation from the signal. In this case, $\bar{n}$ is the signal and $\bar{n}^{1/2}$ is the noise and the signal-to-noise ratio is

$$S/\mathfrak{N} = \bar{n}/\bar{n}^{1/2} = \bar{n}^{1/2} \tag{6.16}$$

If we now reconsider the gray square on the blackboard grid in terms of random placement of the white spots, the situation is not as direct. Whereas before we arrived at a figure of $100\,N - 1$ white spots to locate and shade a gray spot to 1% of the average background, this was done with precise placement of the spots. If the spots are put on at random and the average background is to have 100 white spots in each grid element, the rms deviation is $(100)^{1/2} = 10$. Thus, the noise level is 10 while the signal, which is the deviation of the intensity of the gray element from the average white, is 1. The signal-to-noise ratio is 0.1, a value much too low to detect the gray element. To increase this ratio to 1, a generally accepted value in radio communication for signal detection, the average number of spots in each grid element must be increased. Since the required signal has a 1% deviation from the surroundings, it requires the fraction 0.99 average spots of the surroundings. Thus, the signal is $\bar{n} - 0.99\bar{n}$, while the noise is $\bar{n}^{1/2}$. For this ratio to be equal to 1

$$S/\mathfrak{N} = 1 = (\bar{n} - 0.99\bar{n})/\bar{n}^{1/2}$$

$$\bar{n}^{1/2} = 1/(1 - 0.99) = 1/0.01$$

$$\bar{n} = 10^4$$

Thus, the number of spots, or photons in the present case, must be increased by still another factor of 100 to account for the random placement of the spots.

There is still another factor to be considered, and that is called a confidence level. This is the reasonable assurance that an observed signal is not a false alarm. Clearly, the evolutionary survival of an organism must be related to a satisfactory confidence level of incoming visual signals. This confidence level is the appropriate signal-to-noise level, which we have called K; thus $S/\mathfrak{N} = K$. Although it was suggested above that $K = 1$ for radio signal levels, these are not necessarily satisfactory visual signal levels.

We saw from Eq. (6.15) that the average value $\bar{n}$ is equal to the average value of $(n_i - \bar{n})^2$ or

$$\bar{n} = \overline{(n_i - n)^2}$$

and, since from Eq. (6.16)

$$S/\mathfrak{N} \equiv K = \bar{n}^{1/2}$$

$$K^2 = \overline{(n_i - \bar{n})^2} \tag{6.17}$$

A property of the Poisson distribution function is that it yields a Gaussian probability distribution curve, as in Fig. 6.28. (See for example, Villars and Benedek (1974)). In this figure, P is the Poisson probability density and K is the rms deviation. Since a Gaussian curve is normalized, its area is unity. For example, the area under the curve between $K = 1$ and $K = 2$ is 0.13. This is the probability that an observation lies in the range

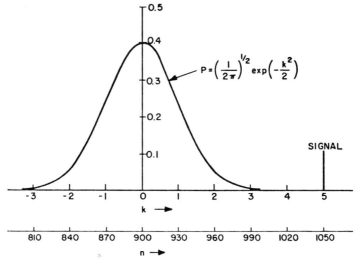

FIG. 6.28 Plot of Poisson probability distribution of a noisy quantity about its mean value. K, rms deviation from the average value which is the signal-to-noise ratio; n, number of counts. Note that for large n the shape of the curve is Gaussian. [Reprinted with permission from Rose (1973). Copyrighted by Plenum Publishing Corporation.]

between 1 and 2 rms deviation above the average signal. The area under the curve above $K = 2$ is 0.023, and therefore this is the probability that an observation will exceed 2 rms deviation. Table 6.4 lists the areas under the curve for various values of K, and therefore it is a table of probabilities of a signal exceeding a given rms deviation.

TABLE 6.4[a]

Values for the Probability of Exceeding Various Values of K

K	Probability of exceeding K
1	0.15
2	0.023
3	1.3×10^{-3}
4	3×10^{-5}
5	3×10^{-7}
6	2×10^{-9}

[a] Reprinted with permission from Rose (1973). Copyrighted by Plenum Publishing Corporation.

Suppose in our picture, as in a typical television tube, we have 10^5 spots. This give 10^5 opportunities to generate a false alarm, i.e., an error in 10^{-5} can give a false alarm. We therefore require a signal-to-noise ratio K less than this value, and in Table 6.4 it is seen that K must be greater than 4. If, for example, the picture had only 10^3 picture elements, a K slightly greater than 3 would be satisfactory. Since the signal itself is noisy, it is safer to choose the next largest K value for the ratio. Therefore, since from Eq. (6.16) the ratio of rms deviation to the average background is $\bar{n}^{1/2}/\bar{n} = 1/\bar{n}^{1/2}$, it is necessary to increase the number of photons by $\bar{n}$, or K^2, in order to decrease the ratio of rms deviation to background by K.

Let us now generalize these terms. In the above example, we wished to have a contrast between the shaded spot and the background at 1% so we wrote the contrast as $\bar{n} - 0.99\bar{n}$, where the background is $\bar{n}$. Let C represent the contrast, or the ratio of change of background to background, i.e., $C = (\bar{n} - 0.99\bar{n})/\bar{n}$. In this case, $C = 0.01$ means 1% contrast, but in general any fraction can be used. Let N be the number of picture elements, which are photon excitations, and K the signal-to-noise ratio, with K^2 required as discussed above to avoid false alarms. The general expression is then

$$\text{total number of photons} = NK^2/C^2 \qquad (6.18)$$

Note that the number of picture elements N is equal to the area of the screen A divided by the area of a spot a and, if the spot is square with a

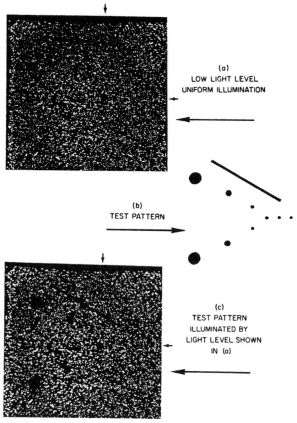

FIG. 6.29 Demonstration of the limitations imposed by the quantum nature of light on its ability to transmit information. [Reprinted with permission from Rose (1973). Copyrighted by Plenum Publishing Corporation.]

width d, then $a = d^2$ and $N = A/d^2$. We can then write

$$\text{total number of photons} = AK^2/d^2C^2 \qquad (6.19)$$

which shows that for a fixed number of photons the width of a detectable object varies inversely with the contrast desired, i.e., if C increases d decreases.

An elementary example of threshold visibility is given by Rose (1973). In Fig. 6.29, each photon is made visible on a television screen with a high gain photomultiplier, and the discreteness of the photons is pictured by the random distribution and consequent noisiness. If we are looking for a black spot of a given size, we are aware of the false alarms in the picture. The test pattern of Fig. 6.29b has been placed over Fig. 6.29a to produce Fig. 6.29c. The four largest black dots can be seen but they appear to terminate at a fifth black dot at the apex of the triangle. This fifth black dot is not present in the pattern but is part of the noise in Fig. 6.29. This constitutes a "false alarm." Let us consider the size of the false alarms on the basis of the K value of the signal-to-noise ratio.

The statistically generated fifth black spot of Fig. 6.29 occupies an area of about $\frac{1}{500}$ of the picture area. There are 4500 dots in the picture so the average number of dots in this area, $\bar{n}$, is 9. From Eq. (6.16) $S/\mathfrak{N} = \bar{n}^{1/2}$ and therefore $K = 9^{1/2} = 3$, and from Table 6.4 it is seen that $K = 3$ will yield false alarms about once in 1000 times. If the picture is subdivided into areas, the size of this spot there would be 500 areas. With a probability of 10^{-3} that one of these areas would be totally black, one would need two screens generated in this fashion to find a single black spot of this size. This is at the threshold of visibility of spots this size. A larger spot would have a correspondingly larger K. For example, the largest black spot of Fig. 6.29b covers an area occupied by about $\bar{n} = 25$ white dots. In this case $K = \sqrt{25} = 5$ and reference to Table 6.4 shows that the probability of false alarms is only 3×10^{-7}. Thus, the large black spots are clearly visible essentially all of the time.

On the basis of studies of this type, the television tube has been designed to have about 10^5 picture elements and a K value of 4–5.

THE EYE OF *LIMULUS*

It is essential to measure the signals induced in the optic nerve by light acting on the retina. In higher forms of life these signals are mediated in the retina by a variety of nerve cells, which will be discussed later, and it is difficult to ascertain a direct correspondence between the photo absorption and the optic nerve signal. The most extensive data, obtained largely by Hartline and his associates over the past 40 years, has been from the arthropod *Limulus polyphemus*. This is the common horseshoe crab, Fig. 6.30a, which is a very primitive form of life. Fossil evidence indicates very little change over about 200 million years. The eye is in a very primitive state of development and about 1 cm of nerve fiber extends directly from the photoabsorber before it joins other nerves and the optic fiber.

The lateral eye seen in Fig. 6.30b is composed of about 600 segments, called *ommatidia*. Although separate from each other, there are cross-links between the nerves so that there is some cooperative mechanism. These cross-links are severed, however, in the preparation. Figure 6.31 is a horizontal section through the eye showing seven ommatidia, where C is the cornea, CO the cone, and R the *retinula* or light sensitive part. There are from 10 to 20 of these retinula grouped about a central axis much like

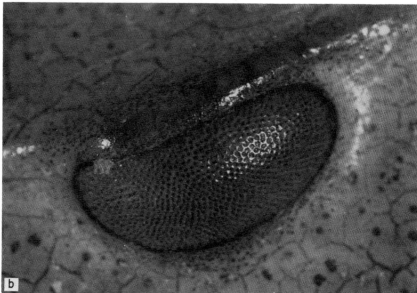

FIG. 6.30 (a) The arthropod *Limulus* commonly called the horseshoe crab. (b) An enlarged view of the eye showing the arrangement of the ommatidia. [From Cornsweet (1973).]

segments in an orange, Fig. 6.32. A longitudinal section is seen in Fig. 6.33. This illustrates the eccentric cell which is a neuron whose upper, or distal, end is connected to the retinular cells and then leads to the optic nerve. As will be seen later, a microelectrode in the eccentric cell is the more sensitive position for electrical measurements.

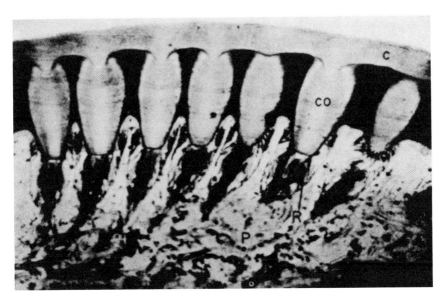

FIG. 6.31 Horizontal section through the eye of *Limulus* showing seven ommatidia: c, cornea; co, cone; R, retinula (light sensitive part). [From Hartline *et al*. (1952).]

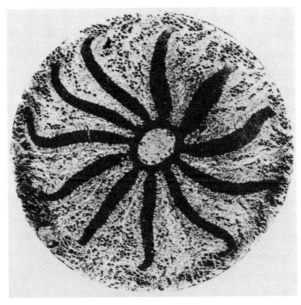

FIG. 6.32 Electron micrograph of the sensory part of *Limulus* ommatidium in cross section. In the center is the distal process of the eccentric cell, with rhabdom shaped like a hub and wheel spokes. [From Ratliff *et al*. (1963).]

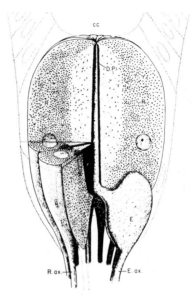

FIG. 6.33 Schematic of *Limulus* ommatidium in longitudinal section. cc, crystal cone; r, rhabdom; R, retinular cell; D. P., distal process; E. ax., eccentric cell axon; R. ax., retinular cell axon. [From Ratliff *et al.* (1963).]

In the experiments a cut is made perpendicular to the surface of the cornea and a nerve strand a little below one of the ommatidia is snipped out. Each never strand contains ten or more fibers. In this way the nerve strand is separated before it joins others in its route to the optic nerve. When the nerve fiber from a single ommatidium is thus isolated, the ommatidium can be separated from the others and placed in a moist chamber. Electrodes are attached at positions 1–4 in Fig. 6.34.

When light is shined on the ommatidium, a slow retinal potential is observed between leads 1 and 2; see upper curves in Fig. 6.35, in which upward deflection means increasing negativity of lead 1 with respect to lead 2. If some of the nerve is included, but the recording is made from leads 3 or 4, a spike action potential is seen, lower traces, Fig. 6.35. This is similar to those seen in nerve axons of other types of sensory receptors. In this figure the lower time markers are 0.02 sec and the black line indicates when the illumination is on. The three different intensities of relative values 1.0, 0.1, and 0.01 from top to bottom show three effects with increasing intensity; (1) the magnitude of the slow rise potential (upper curves) increases, (2) the frequency of the spikes in the axon increases, and (3) the delay time from the onset of illumination to both the beginning of the upper curve and the spikes decreases.

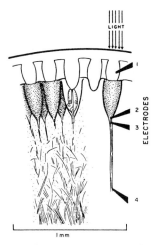

FIG. 6.34 Schematic drawing, representing a section of lateral eye of *Limulus* in a plane perpendicular to surface of cornea, as seen in fresh preparations. Transparent cornea at top, showing crystalline cones of the ommatidia; the heavily melanin-pigmented conical bodies of these form a layer on the inner surface of the cornea. On the left, a group of ommatidia is represented, with indications of bundles of nerve fibers traversing the plexus behind the ommatidia, collecting in larger bundles that become the optic nerve still farther back. One of these ommatidia has been represented as if the section had passed through it, revealing the sensory component, also as if sectioned. On the right an ommatidium with its nerve fiber bundle is represented as it appears after having been isolated by dissection and suspended, in air, on electrodes (moist cotton wicks, from chlorided silver tubes filled with seawater) represented by solid black triangles. [From Hartline *et al.* (1952).]

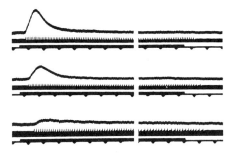

FIG. 6.35 Simultaneously recorded nerve and "retinal" potentials of an isolated ommatidium from the eye of *Limulus*, in response to illumination at three intensities of relative value (top to bottom) 1.0, 0.1, 0.01. Upper trace in each record: potential of the body of the ommatidium (leads 1–2, cf. Fig. 6.34), dc amplification. Lower trace (black edge) in each record; spike action potentials of the nerve strand from the ommatidium. For both traces, deflection upward indicates increasing negativity of distal leads (1, 3) with respect to proximal leads (2, 4). Peak deflection of upper trace in top record = 0.4 mv. In each record, signal marking period of illumination blackens lower half of white band above time marks. Time marked in 1/5 sec. [From Hartline *et al.* (1952).]

COMPARISON OF *LIMULUS* AND THE HUMAN EYE

Analysis of the chemical photopigment in the *Limulus* eye shows that it is rhodopsin as in the human eye. One would therefore expect the same spectral sensitivity. Graham and Hartline (1935), in a similar preparation to that described above, were able to isolate single fibers coming out of an ommatidium. By adjusting the intensity of light for different wavelengths, they could obtain spike potentials in the axon which had essentially the same time pattern for the first three spikes. Some examples are seen in Fig. 6.36. On the left in this figure are listed the wavelengths and relative intensities of 0.04-sec flashes on a dark-adapted eye. Data of this type, normalized so that maximum sensitivity equals 100%, are shown in Fig. 6.37 by the solid curve. The dashed curve is that of a human dark-adapted eye based on statistical measurements of just visible flashes by Hecht and Williams (1922).

The intensity–duration relationship of the *Limulus* eye is essentially the same as that of the human eye (see Figs. 6.16 and 6.17). In Fig. 6.38 a combination of experiments on the *Limulus* eye is seen. The durations are in seconds and the intensity is in arbitrary units. The time markings are 0.02 sec and the black section which interrupts the white line indicates when the light is on. It is seen that the diagonals from upper left to lower right have essentially the same pattern of spikes. This means that the product of intensity I and duration τ in a given diagonal of measurements is a constant E or

$$I\tau = E$$

Clearly, E is energy. This is known as the Bunsen–Roscoe law of photochemical reactions in which the energy delivered is the controlling factor. It is applicable in the eye up to about 0.1 sec. One would not expect exposures greater than 0.1 sec duration to follow this law from an examination of Fig. 6.38. It is seen that the spikes up to 0.1 sec duration begin after the light flash is off; for the 0.1 sec duration the spikes have begun, while the energy from the flash is still accumulating. Clearly, one cannot compare events that take place during the illumination. This is not a failure of the Bunsen–Roscoe law but simply a region of inapplicability.

Dark adaptation also follows the same rate constants as the human eye. This can be seen by the number of spikes from a light flash increasing with time in darkness in Fig. 6.39. These experiments show that the receptor mechanism of the human eye was fully developed in lower animals at least 200 million years ago and has undergone little change since. They further indicate that experiments of this type on *Limulus* are reasonably representative of the behavior of the human eye.

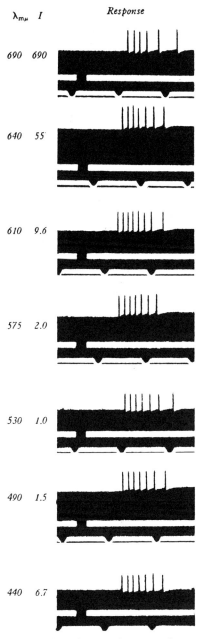

FIG. 6.36 The response of single optic nerve from *Limulus* eye at different wavelengths (left column). The intensity (second column) was adjusted so that the spacings of the first three action potential spikes were the same. [From Graham and Hartline (1935).]

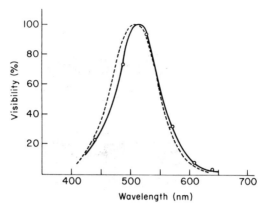

FIG. 6.37 Average visibility curve for the *Limulus* eye (—0—) compared with the human dim vision visibility curve (---). The visibilities are expressed on the basis of maximum visibility equals unity. [From Graham and Hartline (1935).]

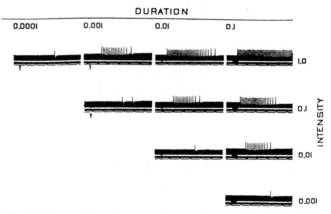

FIG. 6.38 Discharges of impulses in a single optic nerve fiber (*Limulus*) in response to short flashes of light of various intensitites and durations. Relative intensity for each horizontal row given on right. Duration of flash (in seconds) for each vertical column given at top. Signal of light flash blackens the white line above time marker (arrows mark position of signal for very short flashes). Time marks at bottom are in 0.5-sec intervals. [From Hartline (1934).]

INHIBITORY INTERACTION OF RECEPTOR UNITS

The ommatidia of the *Limulus* eye serve as individual receptor units. In the experiments described previously, about 1 cm of nerve fiber from an ommatidium was separated from the other fibers for measurement. Recordings of the action potential can be made equally well by inserting a microelectrode directly into the eccentric cell of an ommatidium. In this

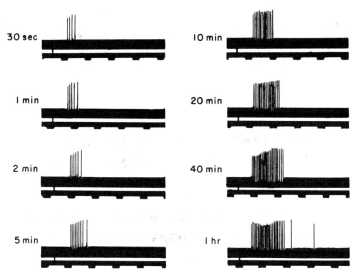

FIG. 6.39 Dark adaptation of a single visual receptor in the eye of *Limulus*. Oscillograms of the amplified action potentials of a single optic nerve fiber, showing the discharge of impulses in response to a test flash of light of fixed intensity applied to the eye at various times in the dark (given at the left of each record) following a period of light adaptation. In each record, deflections of the upper black edge are the amplified action potential spikes of a single active fiber in a small bundle dissected from the rest of the optic nerve. On the lower black edge are time marks ($\frac{1}{5}$ sec); the white band just above contains the signal of the test flash (narrow black stripe near the left–hand edge; flash duration: 0.008 sec). [From Hartline and McDonald (1947).]

way, the effects of maintaining intact the plexus of cross-linked fibers may be determined.

It was found that the *Limulus* eye is not a group of individual receptors but that every receptor is connected to every other receptor with a resulting interaction. The nature of this interaction is that of inhibition, that is, light shining on one ommatidium decreases the response to light of all of the other ommatidia, the inhibitory effect weakening with distance from the illuminated ommatidium. Conversely, illumination of regions in the vicinity of any particular ommatidium reduces the ability of that receptor unit to discharge impulses in response to light. This behavior in the *Limulus* eye was shown by Hartline *et al.* (1956). An example of this is shown in Fig. 6.40. The illumination of a single ommatidium has begun to the left of the figure. Where the white bar ends, a different region of the eye was additionally illuminated. It is seen that there is a noticeable decrease in the frequency of the action potential during the illumination of the surrounding region and when this illumination is turned off, while maintaining the continuing illumination of the particular ommatidium being measured, the frequency returns to its original value.

FIG. 6.40   Inhibition of the activity of an ommatidium in the eye of *Limulus* produced by illumination of a nearby retinal region. Oscillogram of action potentials in a single optic nerve fiber. [From Hartline (1949).]

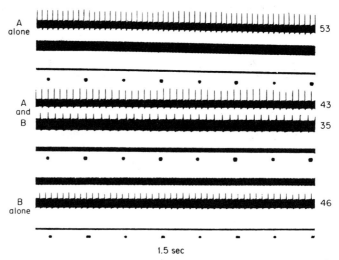

FIG. 6.41   Illustration of inhibition of two *Limulus* ommatidia A and B. Illumination of A alone produces 53 spikes in 1.5 sec. When B is simultaneously illuminated, the firing rate of A is reduced to 43 in 1.5 sec. Similarly with B. [From Hartline and Ratliff (1957).]

Hartline and Ratliff (1957) proceeded from this type of observation to obtain quantitative behavior of inhibition. They measured the discharge rate of the action potential from two nearby ommatidia, called A and B, with two independent sources of light. In Fig. 6.41 it is seen that when A is illuminated alone, the frequency of spikes is 53 in the 1.5-sec record shown, while illumination of B alone elicited 46 spikes in the same time. When both were illuminated simultanesouly, middle record, A was reduced to 43 and B to 35, which showed that each inhibits the other. Data of this type can be used to obtain the plots of Fig. 6.42, which show the decrease in frequency, or inhibition, of each fiber as a function of the frequency of impulses of the other.

The solid lines in Fig. 6.42 were fitted by the method of least squares and their slopes are called inhibitory coefficients K, where K_{AB} is the effect of B on A and K_{BA} the converse. The intercepts on the abscissa are the thresholds of the inhibitory effect and are designated r_A^0 for the effect of A

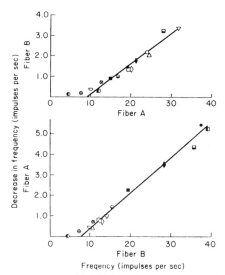

FIG. 6.42   Graphs showing mutual inhibition of two receptor units, in the lateral eye of *Limulus*. In each graph the magnitude of the inhibition of one of the ommatidia is plotted (ordinate) as a function of the degree of concurrent activity of the other (abscissa). The different points were obtained by using various intensities of illumination on ommatidia A and B in various combinations. Frequencies were determined by counting the number of impulse intervals during a fixed time. [From Hartline and Ratliff (1957).]

acting on B and r_B^0 for the effect of B acting on A. If r_A is the response of receptor A to stimulus e_A with no inhibitory effect, then this response is reduced by the illumination of B given by the equation of the line in Fig. 6.42, or

$$r_A = e_A - K_{AB}(r_B - r_B^0) \qquad (6.20)$$

and for B we have an equivalent form

$$r_B = e_B - K_{BA}(r_A - r_A^0)$$

Consider now three receptors, A, B, and C. Since the inhibitory effect decreases with distance of separation A, B and C may be selected spatially so that A and B interact as do B and C, but A and C are too far apart to have much effect on each other. If A is illuminated alone, it will have a certain discharge rate. If now B is illuminated, its inhibitory effect will decrease the discharge rate of A. A similar effect will happen to B if C is illuminated. Now if A is illuminated and then B, A's discharge rate is decreased, but if C is then illuminated, B's discharge rate is decreased and its effect on A is thereby diminished. The effect of C on A through its interaction with B is called *disinhibition*. Thus, one can conclude that no member of the population of receptors is independent of any other member.

These effects on all of the receptors can, in principle, be written as a series of simultaneous equations of the form of Eq. (6.20). However, signals in the cross-links of the nerve fibers have not been measured, so the law of diminution of inhibition with distance is not known nor is it known to be uniform in all directions. In fact, it probably is not. Furthermore, the fundamental unit of reception may not even be a single ommatidium, but instead it may be groups of them connected together.

Inhibition involves the rejection of some of the incoming stimulus by the organism. With so many natural losses already considered in the section on sensitivity of the eye, it may at first seem that this is an evolutionary error. However, it makes sense when the usefulness of information is considered. Inhibition enhances contrast. Consider an array of ommatidia with an edge of two different intensities of light shining on it. The ones in the brighter light inhibit those on the darker side, while the effect of those on the darker side is negligible on the ones in the brighter illumination. This results in a greater contrast because, although some of the bright side signal is suppressed, there is a greater amount of suppression of signals on the dark side. As seen in Fig. 6.42, this would be particularly true when the frequency of discharge of the dark side is below the threshold of inhibition.

The inhibition effect has been considered in terms of electrical network theory (Ratliff and Hartline, 1974). The model is based on filters which cause the high frequencies of discharge (the excited component) to be cut off while also cutting off the low frequencies. Inhibition is therefore like a bandpass filter which tunes the eye to send action potentials to the brain in the intermediate frequency range. Thus, although information already lost cannot be restored, there can be amplification of the remaining information. Inhibition actually increases the amplitude of *change* in the level of illumination at the expense of information about the *absolute* level. The survival value to an organism of increased visual acuity of prey or predator passing by is considerably greater than knowing if it is a sunny or cloudy day. For more extensive discussion of inhibitory effects in neural systems see Ratliff (1965) or von Békésy (1967).

ECCENTRIC CELL MEMBRANE RESISTANCE AND FIRING FREQUENCY

In a careful series of experiments, Fuortes (1959) was able to measure the relationship of the eccentric cell membrane resistance of *Limulus* with changes in illumination and electric potential. He prepared an ommatidium, as described earlier, with a microelectrode inserted in the eccentric cell. With a special bridge circuit, he could change the potential of this electrode, and therefore the cell, and observe the output voltage on an oscilloscope. It should be recalled that the eccentric cell is not a photoreceptor itself, but receives signals from the photoreceptors. Also, as shown in Fig. 6.35, illumination of an ommatidium produces a negative voltage, which may be called a generator potential, that apparently stimulates action potentials, or spikes, in the optic nerve.

When the ommatidium is illuminated, the microelectrode being initially at zero potential, the amplitude of the generator potential and the frequency of firing are both approximately linear functions of the logarithm of the illumination intensity, log I. Therefore, the frequency of firing is a linear function of the generator potential. In this experiment, Fuortes also found that he could produce both the generator potential and the spikes without illumination if he applied a small depolarizing current through the microelectrode to the eccentric cell. Figure 6.43 shows typical data taken from oscilloscope photographs. The ordinate scale is frequency of spike impulses, and the abscissa is the magnitude of the generator potential during illumination (open circles) and the magnitude of the depolarizing current in the dark (closed circles). The slope of the firing frequency to the generator potential is 0.77 imp/sec/mV and that of the frequency to the depolarizing current is 4.44 imp/sec/nA. The ratio of the two slopes is 5.74 mΩ. Similar data on other ommatidia gave a range of slope ratios of 4–9 MΩ. If one assumes that applied currents producing firing at a certain frequency evoke in a cell membrane a depolarization identical with that produced by light, then the above ratio is a measure of the resistance of the cell membrane.

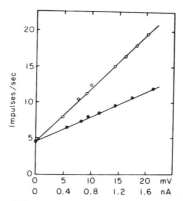

FIG. 6.43 Frequency of firing as a function of generator potential and of current intensity. Abscissa, generator potential amplitude or depolarizing current intensity; ordinate, impulses per sec. Note that the unit was firing continuously owing to some background illumination. O, frequency of firing as a function of generator potential amplitude. ●, frequency of firing as a function of current intensity. [From Fuortes (1959).]

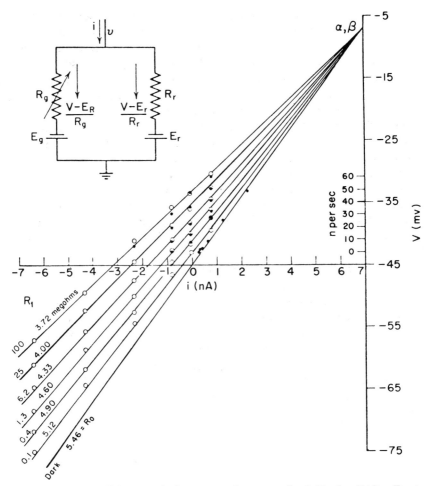

FIG. 6.44 Intracellular records from an optic nerve cell of *Limulus*. [(After Fuortes
(1959).] Abscissa plots depolarizing currents; ordinates plot membrane potential in millivolts
(O) or frequency of impulses per second (●). For a fixed light intensity I (numbers at left of
curves), the circles and dots lie upon the same fixed straight line, whose slope represents the
membrane resistance. The whole structure of concurrent lines follows from the electrical
representation of the membrane shown inset, where R_g is dependent upon light, but R_r is not.
[From Rushton (1959).]

Rushton (1959) performed a more careful analysis of Fuortes' bridge
circuit and, from Fuortes' data, was able to construct the graph of Fig.
6.44. In this figure, the ordinate is the membrane potential (and the firing
frequency) and the abscissa is the depolarizing current. Increasing intensities of illumination are shown on the far left of the figure and the resulting
membrane resistances, the slope of each line, are shown on each line. The

dark resistance is $R_0 = 5.46$ MΩ. The open circles on each line of constant illumination are the membrane potentials and the closed circles the firing frequencies. Thus, as postulated by Fuortes, the membrane potential and firing frequency are linearly related. Thus, along each line of fixed illumination, the membrane resistance is constant and independent of current passing in or out. This behavior is that which would be expected if the cell membrane had the equivalent circuit shown by the insert, where E_r and R_r are the resting potential and resistance and E_g and R_g are the generator potential and resistance, the latter being dependent on illumination intensity.

As noted earlier, the eccentric cell contains no visual pigment so it cannot be a photoreceptor. Since light stimulates the photoreceptors, it might at first be thought that they generate an electric current which in turn stimulates the eccentric cell. But the data of Fig. 6.44 shows that this is not the case, since current alone can excite but that light is required to lower the membrane resistance. Therefore, the role of the photoreceptors must be to generate a substance which lowers membrane resistance.

Another conclusion may be drawn from the data of Fig. 6.44 The frequency of firing (closed circles) fits the same pattern of lines as the membrane potentials (open circles). Thus, they are linearly related to each other and therefore to the membrane resistance ($R_0 - R_1$), where R_0 is the dark resistance and R_I the resistance under illumination. Because frequency of firing is proportional to the logarithm of illumination intensity, a plot of ($R_0 - R_1$) versus log I is functionally the same as a plot of ($R_0 - R_1$) versus firing frequency. Such a plot is shown in Fig. 6.45, where the circles plot R_i versus log I from Fig. 6.44. The curve is the function $\log(1 + I)$ versus log I slid horizontally to fit the points. If it is assumed

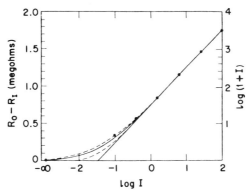

FIG. 6.45 $R_0 - R_1$, the drop in membrane resistance due to light I, is plotted against log I ($\bullet$). ——, the mathematical function $\log(1 + I)$ slid horizontally to fit the points; ---, data from vertebrates. [From Rushton (1959).]

that there is always some thermal noise in the receptor that is present even in the dark, we may call it I_0 since it is equivalent in behavior to some constant illumination in the dark. The equation of the smooth curve of Fig. 6.45 can therefore be represented as

$$R_0 - R_I = \log(I_D + I)$$
$$= \log(1 + I/I_D) + \log I_D \qquad (6.21)$$

where $I_D = -1.4$ log units.

To connect these two conclusions, Rushton (1961) suggested the following. Since the change of membrane resistance is produced by the production of some chemical, the fact that the resistance is always proportional to $\log(I_D + I)$ means that this must be the rate of production of the chemical. A rate process of formation and decay can be constructed to have this functional behavior (see Chapter 7), but whether nature obeys it or not remains to be learned by future investigation.

MICROSTRUCTURE OF RODS

Human rods are roughly cylindrical in shape and are about 0.002 mm in diameter and 0.05-mm long. Each rod contains from 500 to 2000 flattened disks. Radioactive tracer studies have shown that the disks are continuously being replenished at the bottom and disposed of at the top (Graymore, 1970). The lifetime of each disk is a few weeks and varies with the species, with new disks being formed in human rods at a rate of about one every 40 min. Interference with this replenishment by hereditary disorders or vitamin A deficiency can result in impaired vision.

The outer segment of the rod, that part which contains the disks, is the light sensing and signal transmitting part, while the inner segment is involved with cell metabolism. Figure 6.46 shows a schematic drawing of a rod and a cone with only a few disks indicated and the spacing exaggerated. It is seen that the structures of rods and cones are quite different. The chromophore layers in the rods are disks separate from one another. In the cone these layers are formed by repeated folding of the plasma membrane.

Figure 6.47 shows a somewhat different schematic of the cross sections of a rod (left) and a cone (right). Note that in the rod the disks are separated from the plasma membrane by cytoplasm, type M, while in the cones the plasma membrane invaginates the receptor so that the photosensitive membrane is connected continuously to the plasma membrane, type P. Note that the great majority of photoreceptors in nature are of type P, e.g., human cones and the photoreceptors in *Limulus*. Only human rods are known to have the separated disk of type M. A first thought is that comparison between the *Limulus* eye and human rods is not a good one, but, as will be seen in a later section, the detailed ionic transport mechanism is believed to be the same in both type M and P.

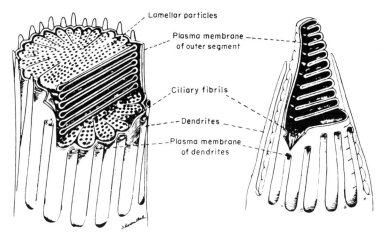

FIG. 6.46 Diagram to show the essential microstructures of *Necturus* rod (left) and cone (right) outer segments. The outer segment of a rod is constructed of a pile of double-membrane disks, each sealed off by a differentiated rim structure. These lamellae are cut radially by a system of deep fissures, and bear a system of deeply staining particles in regular array. The whole is enclosed in a plasma membrane continuous with that of the inner segment. The same plasma membrane is also reflected over the dendrites which stand like a palisade around the outer segment, one in the mouth of each fissure. Running up the outer segment is the residue of the primitive cilium from which the outer segment is derived embryonically. In the cone, the lamellae are formed differently, by the repeated infolding of the plasma membrane. No fissures are present, nor do the lamellae display the deeply staining particles characteristic of *Necturus* rods. [From Wald *et al.* (1963).]

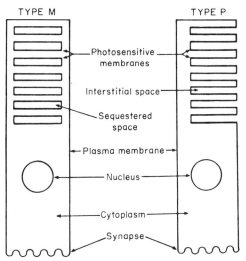

FIG. 6.47 Schematic of the cell membranes of rods (left, type M) and cones (right, type P). [From Hagins (1972). Reproduced with permission from the *Annual Review of Biophysics and Bioengineering*, Volume 1. © 1972 by Annual Reviews Inc.]

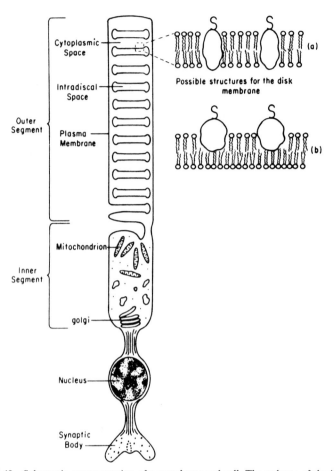

FIG. 6.48   Schematic representation of a vertebrate rod cell. The volume of the intradiscal space is greatly exaggerated, and in the normal bovine outer segment there are approximately 1500 disks within an outer segment 50 μm in length. In the magnified views showing possible structures for the disk membrane, the wiggly lines represent the hydorcarbon chains of the phospholipids, and the open circles the polar head groups. The "S" on the rhodopsin molecule signifies the carbohydrate moiety. [Reprinted with permission from W. L. Hubbell, *Accounts Chem. Res.* **8**, 85 (1975). Copyright 1975 by the American Chemical Society.]

The stacking periodicity of the disks in the rods is about 30 nm. Each disk is about 15 nm in width, and the spacing between two membranes within a disk is about 1.5–2 nm (Korenbrot *et al.* 1973). Therefore each membrane of a disk is about 6.5-nm thick. Other than rhodopsin, the predominant molecule species of the disk membrane is a phospholipid. The phospholipids are arrayed in a smectic liquid crystal phase (layer structure), Fig. 6.48, like other biological membranes (Hubbell 1975), and the rhodopsin, comprising 85% of the dry weight, is an integral part of the membrane. Diffraction experiments have indicated that the rhodopsin molecule behaves as if it were in a planar fluid, and there is experimental evidence that rhodopsin diffuses both rotationally about an axis to the normal membrane and translationally in the membrane. However, the location and distribution of the rhodopsin is still not known. Freeze fracture electron micrographs show both that rhodopsin remains with the outer half of the membrane after fracture and that there are holes on the inner half as if the rhodopsin had penetrated. The inserts of Fig. 6.48 show these two possibilities, where the wiggly lines represent the carbohydrate chains of the phospholipids and the circles their polar heads. The large shapes represent the rhodopsin molecules with the s-shapes being carbo-hydrates. Note from Fig. 6.5 that the photoreceptor part of the rod is at the back of the retina away from the entering light. That is, in Fig. 6.48 the cornea of the eye is in the direction of the bottom of the page.

The disks in Fig. 6.48 are shown as being isolated from the plasma membrane. This isolation is also electrical since cytoplasm acts as an insulator. Thus, any photon which causes a change in the rhodopsin must somehow have its effect communicated to the plasma membrane. Such communication must be chemical or ionic in nature and will be discussed later.

EFFECTS OF LIGHT ON RHODOPSIN

We have discussed earlier that when a quantum of light is absorbed by rhodopsin, the 11-cis isomer of retinal changes to the all-trans isomer and then separates from the opsin molecule. This is the final step of a series of changes that occur. The visual pigment rhodopsin, because it absorbs light so well in the visible spectrum, provides a built-in probe for studying the intermediate changes. This is done by subjecting the pigment, extracted from the rods, to a rapid flash of light and then observing its subsequent absorption spectrum. From such studies, information of the intermediate molecular changes can be obtained which can lead to interpretation of the various configurations. Many of these dynamic changes occur extremely rapidly, and can be slowed only by cooling to low temperatures and even then techniques are required which measure events that occur in nanose-conds.

We have seen in Fig. 6.23 the labeling of two intermediate steps, lumirhodopsin, which is sufficiently stable below $-45°C$ to study, and metarhodopsin, which is sufficiently stable below $-20°C$. This figure also shows their possible configurations and roles. At body temperature these changes take place in about 1 msec. More detailed studies are now available (Abrahamson, 1975).

In a later study, rhodopsin was cooled to $4°K$, excited at 530 nm, and the subsequent absorption spectra measured with picosecond spectroscopy

(Busch *et al.*, 1972). An absorption peak found at 570 nm was assigned to an intermediate now called bathorhodopsin (formerly called prelumirhodopsin). Two significant features found were (1) the formation lifetime was less than 6 picosec and (2) there was very little temperature dependence of this rate of formation between 4° and 77°K. This speed of formation is much too fast from simple mechanical considerations for the isomerization of a large molecule and, even if it were possible, such an isomerization would have a significant temperature dependence. When deuterium was substituted for the hydrogens, there was a large rate change. Because hydrogens contribute only a small amount of the mass of the molecule, the isomerization rate should not be significantly affected. From the kinetics of their data, these investigators concluded that this stage arises from quantum tunneling of a proton, probably that in the Schiff base, to a different position (see Rentzepis 1978).

Bathorhodopsin (prelumirhodopsin) is relatively stable to about −140°C, whereupon it converts thermally into lumirhodopsin. This in turn changes to metarhodopsin I above −40°C, which above −15°C is in chemical equilibrium with metarhodopsin II. This equilibrium is shifted toward metarhodopsin II by acid, and it has been shown that there is an associated uptake of a hydrogen ion while phospholipid is released. The large activation energy and entropy changes associated with this suggest that there are substantial changes occurring in the lipoprotein configuration of the membrane. This process is both rapid and requires an aqueous environment for the proton uptake, so it appears to be the key intermediate step in the entire process that takes place at biological temperatures.

A summary graph of decay time (relaxation time) of the various intermediates which have been distinctly identified by characteristic absoprtion

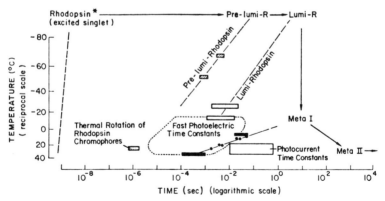

FIG. 6.49 An "Arrhenius map" of some events known to take place when a photon is absorbed in a vertebrate photoreceptor. [From Hagins (1972). Reproduced by permission from the *Annual Review of Biophysics and Bioengineering*, Volume 1. © 1972 by Annual Reviews, Inc.]

spectra is shown in Fig. 6.49. A singly activated decay time τ which is temperature dependent can be written as

$$\tau = \tau_0 \exp(E/RT) \tag{6.22}$$

where τ_0 is a constant, E the barrier height, R the gas constant, and T the absolute temperature. This is known as the Arrhenius equation.[†] Thus, a plot of $\ln \tau$ versus $1/T$ yields a straight line whose slope $\Delta G/R$ is proportional to the activation energy ΔG. In Fig. 6.49 the ordinate is reciprocal absolute temperature corrected to degrees Centigrade for ease of relating to biological processes. The fastest reaction, excited singlet, is expressed as a dashed line because no data exist, and it is doubtful if there is a thermal activation for such processes. Prelumi- (now called batho-) and lumirhodopsin reactions are so fast that their relaxation times at physiological temperatures can only be estimated from low temperature data. Metarhodopsin I is the intermediate which happens to be associated with the visual process because it is the last one to disappear before the first detectable phototransduction in the retina. This occurs in a few milliseconds, and therefore metarhodopsin II is too slow. The solid bars give the approximate range of decay times of metarhodopsin I in solution, which may not be quite the same as *in vivo*, and the solid circles represent decay time constants for the excitatory process in rabbit eyes.

Attempts at analysis of the possible chemical kinetics and their comparison with the time course of the transducer current have been made (Fuortes and Hodgkin 1964). The electrical circuit model was that of filters in series. Parameters were varied to match the measured waveforms of electrical responses. It was found, however, that the minimum number of filters, each of which represents a chemical reaction with its rate constant, is greater than the known number of chemical intermediates. Furthermore, none of the electrical time constants match any of the relaxation times of

---

[†]The Arrhenius equation can be obtained from the following simple argument. For a system in which there is no pressure or volume change in a reaction, the Gibb's free energy (Vol. 1, Appendix Section A.1), $\Delta G = \Delta U + \Delta(PV) - \Delta(TS)$, reduces to the Helmholz free energy, $\Delta F = \Delta U - \Delta(TS)$, where ΔU in reference to some standard state U_0 is the thermal energy E. If there is an entropy change required in addition to thermal energy for an atom to surmount a barrier of height E, then the Boltzmann probability for jumping over the barrier at a given temperature T is $e^{-\Delta F/RT} = e^{-E/RT}e^{\Delta S/R}$, where ΔS is the entropy change associated with the rearrangement of molecules and changes in vibrational frequencies. If there is minimal entropy change, $e^{\Delta S/R}$ is about unity. Thus, the frequency of atoms jumping over the barrier is the number of attempts per second times the above probability that an atom will have sufficient thermal energy E to surmount the barrier. The attempt frequency is taken to be the vibrational frequency ν, which is of the order of $10^{14}/\text{sec}$. Thus, the jump frequency Γ is given by $\Gamma = \nu e^{\Delta S/R} e^{-E/RT}$ or $\Gamma = \Gamma_0 e^{-E/RT}$, where $\Gamma_0 = \nu e^{\Delta S/R}$. The mean time between jumps τ is the reciprocal of the jump frequency or $\tau = \tau_0 e^{E/RT}$, where $\tau_0 = 1/\Gamma_0$. [For further discussion see Hinshelwood (1945).]

the known chemical intermediates. Further work along this line has been done by Borsellino and Fuortes (1968), who have been able to reproduce many features of the transducer kinetics at low light intensities. However, the assignment of the kinetic steps to discrete chemical stages has not been accomplished.

THE TRANSDUCTION PROCESS

When a flash of light is absorbed by rhodopsin in rod cells, the first electrical excitation of neurons in the retina can be detected within the order of 1 msec. We have seen earlier that the retina is sensitive to a few quanta, or at most one quantum is required to initiate the transduction process in a rod. This initiating quantum must have an amplification factor of several orders of magnitude to cause the observed change in membrane potential. This potential develops and then decreases over a few hundred milliseconds. This potential, as will be shown in the next section, causes changes in potential in higher order neurons in the retina, and is finally encoded into an all-or-nothing spike potential which is transmitted through the optic nerve to the brain. How does a single event from a single quantum accomplish this?

The initiating event has been a subject of considerable study, and Yoshikami and Hagins (1971) proposed a model which has received considerable attention. Hagins (1972) has elaborated on the model and we shall briefly discuss it.

Tracer studies on vertebrate rods using radioactive Na^+ ions have shown that there is a very large current of Na^+ ions flowing inward through the plasma membrane in the dark. This dark current is so large that it has been estimated to be equivalent to a replacement of all of the cations in a rod in less than 1 min. If sodium is absent from the external medium, the current stops. Tracer evidence also indicates that an equivalent amount of sodium is being continuously ejected by a sodium pump. In light, this Na^+ current is considerably reduced. It was further found that the dark currents are largest in solution with 10^{-5} mole Ca^{2+} ions and decreased as the calcium ion concentration was increased until they eventually disappeared. The effects of external calcium ion concentration are rapid and reversible. Based on these findings, the hypothesis is that a calcium pump exists in the disk membrane which maintains a higher concentration within the membrane than outside. Light increases the disk membrane permeability to calcium so that it is released to block the sites through which Na^+ ions have been entering the cytoplasm from the outside. Because of the resulting deficiency of Na^+ inward flow, the interior becomes more negative and it is this hyperpolarization of the plasma membrane close to the affected disk which is transmitted. The amplification comes about because in some way the 11-cis to all-trans configuration opens channels within the disk to release calcium ions into the cytoplasm. These ions then diffuse to the Na^+ channels in the plasma membrane and block them. This hypothesis is illustrated schematically in Fig. 6.50 for a rod disk (left) and a cone segment (right), the same mechanism applying to both. The upper figure shows the calcium pump maintaining a high calcium ion concentration on the side of the chromophore membrane away from the cytoplasm. The Na^+ dark current channels are shown but not the Na^+ pump channels which eject sodium to maintain ionic balance. In the lower figure, light has caused the Ca^+ ions to leak into the cytoplasm and block the Na^+ dark current channels.

Recent studies are testing this hypothesis, particularly the role of Ca^{2+} ions. Mason and Lee (1973) found that membranes can reseal themselves. These investigators broke up cell suspensions, rod disks as well as red blood cells, with an ultrasonic device. This process is called sonication. Within 1 hr, the fragments of ragged cell walls had formed themselves into small vesicles, sealed and with smooth walls. Mason *et al.* (1974) made

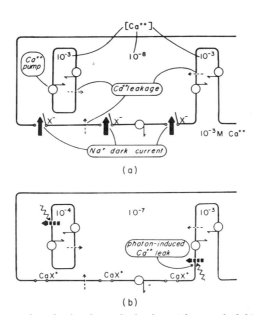

FIG. 6.50 A proposed mechanism for excitation in vertebrate rods (left) and cones (right). Values for calcium ion activities in cytoplasm and disks are estimates derived from studies of nerve and muscle and do not represent actual measurements. (a) Rod/cone in darkness. (b) Rod/cone in light. [From Hagins (1972). Reproduced with permission from the *Annual Review of Biophysics and Bioengineering*, Volume 1. © 1972 by Annual Reviews, Inc.]

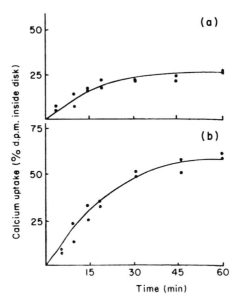

FIG. 6.51 Uptake of calcium by disk membranes (a), in the dark, and (b), in light. Membranes were incubated in 100-mM Tris buffer containing 10^{-5} M Ca^{2+} at 23°C. [Reprinted by permission from W. T. Mason *et al.*, *Nature, New Biol.* **247**, 562 (1974). Copyright © 1974 Macmillan Journals Limited.]

such vesicles from the rod disks of frogs. They incubated these in a 10^{-5} mole Ca^{2+} solution with a $^{45}Ca^{2+}$ tracer for times up to 1 hr. They then removed the vesicles by centrifuging, washed off the solution, and counted their activity. Although the disk membrane occupied not more than 5% of the suspension volume, it removed nearly 50% of the Ca^{2+} from the solution. After the disk vesicle achieved an equilibrium Ca^{2+} concentration in its interior, the remaining calcium had to enter against a gradient. This clearly suggests the presence of a Ca^{2+} pump in the membrane. The data are shown in Fig. 6.51, where the ordinate is a count of the disintegrations per minute of the radioactive calcium. They further found that after exposure to light, which bleached the membranes, subsequent incubation in the dark resulted in a much faster uptake of Ca^{2+}; compare the lower curve, bleached, with the upper curve, unbleached, in Fig. 6.51.

Having demonstrated that the Ca^{2+} pump exists and that the action of light causes a faster uptake, Mason *et al.* then examined the reverse effect. Does light cause the ejection of Ca^{2+} from the membrane vesicle? For this experiment, they put the Ca^{2+} initially inside the vesicles by sonication and resealing in a 10^{-4} mole solution of Ca^{2+} with radioactive $^{45}Ca^{2+}$. They then transferred these to a neutral solution which contained a chemical to prevent Ca^{2+} from being pumped back in. They exposed some

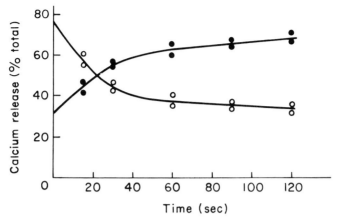

FIG. 6.52 Light-induced calcium release by sonicated bovine photoreceptor membranes over a time course of 120-sec, bleaching. Filled circles indicate percentage of Ca^{2+} in the extradisk space and open circles the percentage of Ca^{2+} in the intradisk space. [Replotted from Mason *et al.* (1974).]

vesicles to light and counted the activity remaining in the vesicles and in the external solution. They did this for various intervals ranging from 0 to 120 sec. The 120-sec exposure was selected because absorption spectra showed the rhodopsin to be completely bleached after this time. Figure 6.52 shows the results for the 120-sec exposure. The lower curve shows the decrease of Ca^{2+} remaining in the vesicles and the upper curve the increase in Ca^{2+} outside the vesicles. These results are in agreement with that part of the Yoshikami–Hagins model which hypothesizes the ejection of Ca^{2+} that had been pumped in. Mason *et al.* further showed that the amount of Ca^{2+} released is stepwise in response to bleaching. That is, for a certain amount of bleaching, a certain amount of Ca^{2+} is released. This ratio is about unity in that a mole of Ca^{2+} is released for a mole of bleached rhodopsin.

Thus, three important properties of disk membranes have been demonstrated by Mason *et al.* They concluded that: (1) the disk membranes can actively accumulate Ca^{2+} in the dark and light, (2) the rate of accumulation is nearly three times greater in light than in dark, and (3) the disk membranes release Ca^{2+} as a consequence of light absorption and visual pigment bleaching. Reviews of subsequent experiments on the role of calcium are given in Barlow and Fatt (1977). This specific role of Ca^{2+} is undergoing intensive investigation with some agreement (Wormington and Cone, 1978), some disagreement (Bertrand *et al.*, 1978; Flaming and Brown, 1979), or that calcium is a co-factor (Arden and Low, 1978).

STRUCTURE OF THE PRIMATE RETINA

There are over 100 million primary receptors in the human eye, about 95% of which are rods and the remainder cones. These make connections through a variety of cells within the retina to ganglion cells which send only about 1 million optic nerve fibers to the brain. Thus, the brain receives only a convergence of visual information, most of the integration being performed within the cells of the retina.

The cells of the retina are arranged in five layers, and detailed microscope studies are best interpreted by means of a schematic. Figure 6.53 shows such a drawing. Note that, as mentioned earlier, light passes through the neural cells to reach the chromophore, i.e., from the bottom of the page toward the top. It is seen that the pathway between the receptors and the ganglion cells is much more complicated than a simple line through the

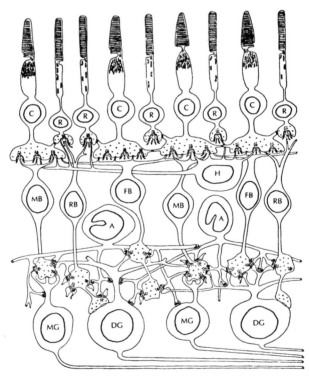

FIG. 6.53 Semischematic diagram of the connections among neural elements in the primate retina that were identified as of 1966. R, rod; C, cone; MB, midget bipolar nerve cell; RB, rod bipolar; FB, flat bipolar; H, horizontal cell; A, amacrine cell; MG, midget ganglion; and DG, diffuse ganglion. The regions where the cells are contiguous are synapses. [From Dowling and Boycott (1966).]

various cells; there are different types of cells with cross information from other receptors. This schematic was developed from detailed electron micrographs by Dowling and Boycott (1966). A summary of their description will now be given and, in a later section, their interpretation of the circuitry will be shown.

The two types of receptor rods (R) and cones (C) are drawn to show some of their fine-structure characteristics. The rods end in small outer-segment disks, while the cones end in large ones with three invaginations (indentations). Three synapse processes usually enter the cone invaginations, while between four and seven enter the single rod invagination. There are three different types of bipolar nerve cells: midget bipolar (MB), rod bipolar (RB), and flat bipolar (FB). The rods connect with the rod bipolars and the cones with both the midget and flat bipolars; these two bipolars make separate connections with the cone. The midget bipolars connect with only a single cone but they synapse with that cone three times in the triad, e.g., the left MB cell in Fig. 6.53. The flat bipolars connect with a group of cones but make junctions with the cones only superficially, i.e., their axons do not enter the invaginations as do those of the midget bipolars. The horizontal cells send axons to both rods and cones. Further down, it is seen that midget bipolar cells contact a single midget ganglion (MG) cell, and each midget ganglion synapses several times with its midget bipolar. Diffuse ganglion (DG) cells connect with all types of bipolar cells. Amacrine cells contact all the bipolar terminals, and at every bipolar junction both amacrine and ganglion cell connections are present. At these junctions, the amacrine extensions synapse back to the bipolar cells. Amacrine cells also contact each other.

PROCESSING OF RECEPTOR STIMULI WITHIN THE RETINA

It is experimentally easier to study retina neural processing in some lower vertebrates than with primates. Detailed measurements have been made by Werblin and Dowling (1969) on the retina of the mudpuppy *Necturus maculosus*. The retinal neurons in *Necturus* are about 30 μm in diameter compared with diameters of less than 10 μm for mammals. This ratio of diameters yields a volume about 10 times greater and facilitates the insertion of a probe.

Very fine hollow glass probes were inserted into the retina of decapitated heads and recordings were made with different illumination shapes and intensities. After completing the measurements on a given neuron, a small amount of dye in the probe was injected. Subsequent dissection under a microscope enabled the investigators to identify the type of cell on which the measurements had been made. Two shapes of illumination were used, a

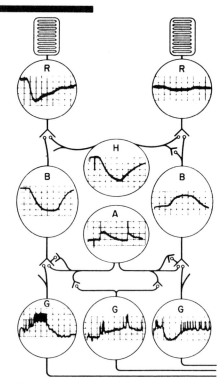

FIG. 6.54   Typical oscilloscope traces of intercellular responses from neurons in *Necturus* with associated synapses. Left, response from a spot of light; right, responses in the surround. Voltage calibration: one scale division equals 1 mV in R; 2 mV in H, B, and G; and 5 mV in A. Time calibration: one division equals 200 msec. R, receptor cells; H, horizontal cells; B, bipolar cells; A, amacrine cells; G, ganglion cells. [From J. E. Dowling, Organization of the vertebrate retina, *Invest. Opth.* **9**, 655 (1970).]

spot and an annulus. The spot was about 100 μm in diameter, and the linewidth of the ring-shaped annulus was also 100 μm. Two diameters of annuli were used, 250 μm and 500 μm. These shapes permitted the estimation of the size of the receptive field, and, when used at the same time as the spot, they showed the antagonistic (inhibiting) effect of surrounding illumination on the signal from a spot illumination. Typical results are shown in Fig. 6.54, in which the left side records responses from a spot of light and the right side the responses from the annulus.

Receptor Cells (R).   When a spot of light is focused on the retina, the receptor cell (R) potential decreases from its typical resting potential of -30 mV by about -4 mV, with a latency between light stimulus and response of about 50 msec and a time to maximum of about 50 msec. Such an increase in negativity is called *hyperpolarization*. (When the resting potential becomes less negative it is called *depolarization*.) Note here that the *Limulus* photoreceptor, which has the cone type of structure depolarizes when exposed to light. The hyperpolarization is graded over a relatively narrow range of light intensities about 2 log units above threshold. At higher intensities the latent time to begin hyperpolarization and the time to reach maximum hyperpolarization both decrease, but the magnitude of the response is unchanged. Very little change is seen for annular illumination. The annular illumination was also turned on while the spot was on and no change was observed. Thus, it can be concluded that the receptive field is 250 μm or smaller in diameter and is unaffected by illumination surrounding the spot. The output from the receptor cell goes into the first synapse indicated in Fig. 6.54.

Horizontal Cells (H).   The horizontal cells hyperpolarize with a latency of about 100 msec and a time to maximum of about 300 msec with a comparably slow recovery time. The magnitude of the response is graded over about 3 log units of intensity above threshold and is dependent on the shape of the stimulus. That is, the response is greater and saturates at a higher intensity when annular rather than spot illumination is used, although the total intensity of the annulus is the same as that of the spot. This suggests that the potential in the horizontal cells is formed by the weighted summation of the potentials from many sites, each of which is saturated at a lower intensity. The response is about the same for a spot of 250 or 500-μm-diameter annulus. The resting potential is about -30 mV, and the response can hyperpolarize the cell by another -20 mV. Microscopic examination indicates that horizontal cell processes extend over distances of 200–400 μm or about the size of the annular illumination. They also show connections with the synapses between the receptor cells and the bipolar cells. This is seen in the schematic of Fig. 6.53. Therefore, Fig. 6.54 indicates that the output from the receptor cell is altered by that of the horizontal cell before the signal reaches the bipolar cell.

Bipolar Cells (B).   The bipolar cells exhibit slow graded responses like the receptor and horizontal cells. However, the experiments indicate that the receptive field of the bipolar cells is organized concentrically into two antagonistic zones. If a spot illumination or annular illumination alone is used, the response of the bipolar cell is to hyperpolarize about -10 mV from a typical resting potential of -30 mV. If spot illumination is used and a 250-μm-diameter annular illumination is superimposed, depolarization can be observed, the amount being proportional to the ratio of the intensities. The right-hand part of Fig. 6.54 indicates the behavior of the bipolar potential with both central and peripheral illumination. Therefore, peripheral illumination appears only to turn off the response of a bipolar cell to central illumination. However, the effect of the peripheral illumination starts about 100 msec later; with a simultaneous spot and annular illumination the bipolar cell first hyperpolarizes and then depolarizes. Microscope studies, as in Fig. 6.53, indicate that the bipolar cell has a spread of about 100 μm. This is about the size of the spot illumination. The 250-μm-diameter annular illumination cannot touch it directly. This is further evidence that the horizontal cells are being affected by the peripheral illumination and connect their response to the first synapse of Fig. 6.54.

Amacrine Cells (A). At this stage of the retina the slow, sustained, graded hyperpolarizing responses are changed to the first type of spike potential in a depolarizing direction. The resting potential is typically -30 to -40 mV and the depolarizing response, including the height of the spike, can be up to -30 mV in magnitude. The spike is superimposed on the beginning of a large, transient, depolarizing curve which may take several hundred milliseconds to decay. Both the magnitude of the spike and the latency, i.e., time from onset of illumination, vary greatly with intensity of illumination. Latencies can be over 600 msec at threshold and less than 200 msec at intensities of 3 log units above threshold. If the intensity changes abruptly the spike is seen, but if it is increased very gradually no spike occurs. It is therefore concluded that amacrine cells respond to sudden changes in illumination which may be related to the observation of motion of an object. The mechanism of conversion of the sustained bipolar response to a transient response across the dipolar-to-amacrine synapse is not yet understood. Werblin and Dowling suggest the following possibility. Figure 6.53 shows that the amacrine cells synapse with both the bipolar cells and the ganglion cells but also there appears to be a return synapse from the amacrine back to the bipolar terminal. If this return synapse were inhibiting, the amacrine might be able to turn off its own excitation.

Ganglion Cells (G). Ganglion cells have a variety of resting potentials never exceeding -40 mV. Their basic response is to depolarize at about the same rate as the bipolar cells. However, they exhibit spike potentials superimposed on the depolarization in three different ways, with a rate of firing roughly proportional to the magnitude of the depolarization. Some respond with illumination on, some with it off, and some with both on and off. Some cells respond with a sustained series of spikes under a spot illumination but this can be inhibited by peripheral illumination. Figure 6.54 shows some examples of these varieties of responses. It also indicates, as does Fig. 6.53, that some ganglion cells can be driven directly by the bipolar cells and some by the amacrine cells. From the variety of responses elicited, it is quite possible that there are more than two types of ganglion cells. In general, ganglion cells are less responsive to simple stationary stimuli than to complicated stimuli. As will be seen in a later section, some respond only to a moving light or shadow edge, while some more specialized ganglia respond only to movement in a given direction. They seem to be finely tuned to certain aspects of motion, and all of this response appears to take place within the retina, with only the results being sent on to the brain through the optic nerve.

THE RECEPTIVE FIELD IN THE MAMMALIAN RETINA

The concept of receptive field in the retina has been discussed in earlier sections. It will be recalled that a single optic nerve fiber of *Limulus* discharged when light fell within a certain small area of the retina, but the discharge could be inhibited by illumination of the retina at a nearby region, the closer to the bright spot the greater the inhibition. In the mammalian retina, which has the complexity of cells described in the previous section, the structure of inhibition (sometimes called antagonistic) is more detailed. Some of the observations of this detail will now be described.

Kuffler (1953) performed a series of experiments on the retina of cats. The eye was fixed in position and a microelectrode was inserted in a ganglion cell. A small beam of light of 0.2-mm diameter was used to explore the region of the retina which caused excitation in the ganglion. First, Kuffler found that a new definition of receptive field was required. While Hartline (1940), in his study of the receptive field in frogs, defined it as the area of the retina which must receive illumination in order to cause discharge in a particular ganglion or nerve fiber, Kuffler enlarged the definition to include all areas which have a functional connection with a ganglion cell. This change in definition was necessary because, although a spot of light of a given intensity in a small area caused a discharge, an increase in intensity of the spot on regions outside the first area could also cause a discharge.

In addition to this, Kuffler found three different types of responses within a receptive field. These are shown in Fig. 6.55. In this experiment a spot of light 0.2-mm diameter was moved to three different regions within a receptive field. In (a) the light flash was near the electrode tip and the ganglion cell discharged when the light was on, although the discharge was not maintained for the duration of the light. In (b) the light spot was 0.5 mm away from the electrode tip. It is seen that no discharge occurred when the light was on but it did occur when the light was turned off. In (c) the light spot was between the regions of (a) and (b), and it is seen that there was discharge both when the light was turned on and when it was

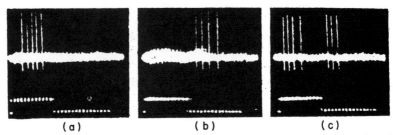

(a)   (b)   (c)

FIG. 6.55 Responses of a ganglion cell in the retina of a cat when a light spot is in three different positions relative to the electrode. (a) Light spot near electrode causes discharge when turned on, (b) light spot 0.5 mm away from electrode causes discharge when turned off, (c) an intermediate position between (a) and (b) causes discharges when turned on or turned off. Impulse height 0.5 mV. [From Kuffler (1953).]

turned off. These three types of responses are called, respectively, "on," "off," and "on–off."

Figure 6.56 shows a typical distribution of these discharge patterns in a receptive field. The ganglion cell is at the tip of the electrode. The clear area in the center is the region of "on" responses, indicated by crosses. The outer area is that of "off" responses, indicated by circles, and the area with horizontal cross-hatching is that of "on–off" responses. Note that the receptive field is not fixed in size or shape but varies with both intensity of illumination and background illumination. Further, the center of a given

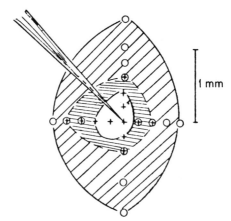

FIG. 6.56 Distribution of discharge patterns within receptive field of ganglion cell (located at tip of electrode). Exploring spot was 0.2 mm in diameter, about 100 times threshold at center of field. In central region (+) "on" discharges were found, while in diagonally hatched part only "off" discharges occurred (O). In intermediary zone (horizontally hatched) discharges were "on–off." Note that change in conditions of illumination (background, etc.) also altered discharge pattern distribution. [From Kuffler (1953).]

receptive field may be either of the "on" type or "off" type, so a detailed mapping of the retina and ganglia is not possible at present. Further details of these and other similar experiments are given in Kuffler and Nicholls (1976).

Other details of the receptive fields have been elucidated. Two of the "on" and "off" centers have been identified; (X-cells which give a sustained discharge and Y-cells which give transient ones (Enroth-Cugell and Robson 1966; Cleland *et al.* 1971), and it has been shown that the axons of the X-ganglion cells conduct more slowly than those of the Y-ganglia. Others have reported ganglion cells with centers (in the receptive field) that respond to both light or dark spots (Stone and Hoffman 1972). These are called W-cells.

It is clear from these types of experiments that the most importamt information conveyed by the ganglion cells is that of sudden changes whether by increase or decrease of brightness. Such changes can come about either by movement of the object or by the rapid scanning (saccadic) movements of the eye. In the next section effects of moving light and shadows will be described.

DIRECTIONAL SELECTION OF MOVEMENT

Experiments to be described were performed by Barlow and Levick (1965) on rabbits. Like the preceding experiments on the cats, the eye was immobilized and a microelectrode was inserted into a ganglion cell. When the associated receptive field was mapped, a small spot of light, or in some cases shadow, was moved across the field. After the direction of motion was found that yielded maximum response, data were taken. It was found that there is a "preferred" direction which yields maximum response in the discharge rate of a given ganglion cell, and that motion in a direction opposite to this, called the "null" direction, yielded little or no response.

Figure 6.57 shows the detailed mapping of a typical receptive field indicated by the circular enclosure at the top of the figure. Also indicated are the preferred and null directions of motion, by the arrows to the left. In the lower trace of each pair of traces downward motion indicates preferred and upward motion the null direction; the corresponding discharge patterns of such motion is shown in the upper traces. In the lower three rows the larger response for the preferred direction is quite evident. The experiments were repeated with small slits of light, the slit width varying from 8 min of arc to 1 deg. These experiments showed that the angular width of the subunit of the receptive field responsible for directional selectivity is about $\frac{1}{4}$ deg.

In another type of measurement two slits of 0.1-deg width were placed

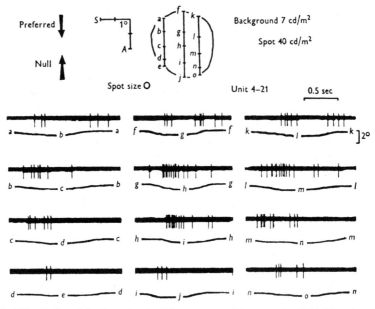

FIG. 6.57   Back and forth motion in different parts of the receptive field. The edge of the receptive field is mapped at the top, and the positions $a, b, c, \ldots, o$, within it are indicated. The spot was moved back and forth several times between a and b, then between b and c, and so on. The records are samples of these back and forth motions. The lower trace of each pair shows the position of the spot in the field: downward movement of the trace corresponds to movement of the spot in the preferred direction. Marked asymmetry of response for the opposite directions holds in most positions in the field. Its absence in the top row of records is expected in the inhibitory scheme. [From Barlow and Levick (1965).]

side by side with varying distances between them; the distance was measured by the angle subtended on the retina. First, the response of each slit was measured independently. A card was moved across each slit, and each gave the largest response when the card was moved in the preferred direction. Then, the slits were illuminated one after the other in both the preferred and the null directions. Figure 6.58 shows results for the 17-min and the 1° 6-min separations. In both cases slit A was arranged to give the bigger step in illumination. Part of the light struck a photomultipler, and the output is shown by the lower trace. The response in the ganglion cell is shown in the upper traces when the slits are illuminated sequentially in the preferred and null directions and when the illumination is turned off sequentially. For the 17-min separation of slits the number of spikes is clearly greater in the preferred sequence but a 1° 6-min separation this distinction is not clearly discernible.

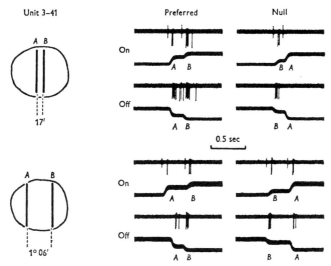

FIG. 6.58 Responses to different temporal sequences of two static stimuli. On the left the positions of the pair of stimuli are shown within the outline of the receptive field. Records for the small separation are shown above, those for the large separation below. Within each half, records for "on" are above those for "off." The lower trace of each pair is the photomultiplier output: increasing light moves the trace upwards, and slit A was arranged to give the bigger step in every case, even though it was not brighter. Preferred sequences are on the left, null on the right. Notice that the preferred sequence yields more spikes than the null at the small spatial separation, but this difference ceases to be clearly visible when the separation of the slits is increased. [From Barlow and Levick (1965).]

If a single spot of light is moved in the null direction no response occurs. However, if the motion of the spot is temporarily stopped, when it starts again, discharge occurs. Furthermore, if the spot is moved slowly across the receptive field in the null direction, there is discharge in the ganglion cell. These experiments suggest that there is either a transit time Δt for the inhibitory signal or that the persistence of the signal has finite duration and, when it ends, there is no further inhibition.

Barlow and Levick's model for this behavior is shown in Fig. 6.59. If light is moving in the preferred direction, from left to right, the signal from receptor A will reach the ganglion before the signal from B can inhibit it, similarly with B. If motion is in the opposite, or null, direction, the inhibitory signal from C will reach the ganglion cell at the same time as does the signal from receptor B and cause cancellation. The horizontal transmission labeled Δt would be functionally performed by the horizontal cells of the retina, see Fig. 6.53.

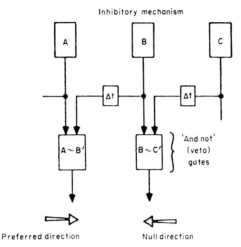

FIG. 6.59 Hypothetical method for discrimination sequence. The preferred direction is from left to right, null from right to left. In the excitatory scheme activity from the groups of receptors A and B is delayed before it is passed laterally in the preferred direction to the "and" (conjunction) gates. In this scheme the activity spreads laterally, but in the null direction, from the groups of receptors B and C, and it has an inhibitory action at the units in the next layer; hence these act as "and not" (veto) gates. The inhibition prevents activity from A and B passing through these gates if motion is in the null direction, but arrives too late to have an effect if motion is in the preferred direction. Notice that a special delay unit is not really necessary, for this scheme works if inhibition simply persists longer than excitation and can thus continue to be effective after a lapse of time. [From Barlow and Levick (1965).]

THE CIRCUITRY OF THE RETINA

The microstructure of the cells of the retina, Fig. 6.53, was described earlier. With the knowledge of receptive fields, we can now appreciate the meaning of the connections of the various cells as deduced by Dowling and Boycott (1966) and others.

It is known that various types of receptors interact principally only with receptors of their own kind, that is rods inhibit only rods. This suggests that the neurons of the retina which connect receptors, horizontal and amacrine cells, connect only receptors of similar type.

As we have seen in Fig. 6.56, the receptive field of the mammalian retina consists of two concentric regions which are antagonistic. Illumination of the center of the receptive field causes the cell to fire when the light either comes on or goes off, while simultaneous illumination of outer ring inhibits the effect. For example, if illumination of the center of the field causes the cell to fire, illumination of the periphery reduces the firing rate. Across the retina the size of the center of the receptive field varies, becoming very small, possibly the size of a single cone, at the fovea. However, the size of the antagonistic part of the receptive field remains essentially unchanged. Furthermore, at the fovea the lateral spreads of many of the bipolar terminals and ganglion connections are considerably smaller than those outside the fovea. This may be correlated with the reduced size of the central region of the receptive field in the fovea. Also, in the fovea there are individual vertical pathways for the cones to the midget bipolar cells and then to the midget ganglion cells, as indicated on the left side of Fig. 6.53. Again, this correlates with the small center of the receptive field in the fovea. On the other hand, the lateral extensions of the horizontal and amacrine cells do not seem to be reduced in the fovea. This strongly suggests that they are responsible for the antagonistic field, which is unchanged in size between the fovea and the periphery. Thus, a model can be developed in which the pathway at the center of the receptive field is receptor-bipolar-ganglion, while the pathway for the antagonistic field is receptor-bipolar-interneurons-(horizontal and amacrine)-ganglion. If the antagonistic field response goes through the interneural cells, there are additional synapses to pass through. It would be expected that there would be a time delay of this response because of this, and indeed it has been observed. A further observation has permitted the separation of roles of the interneural cells, horizontal and amacrine. The periphery of the receptive field of a cat's retina may extend great distances, in fact, response has been shown up to 1 cm from the center of the receptive field, with an appropriate delay time in the signal affecting the response of the center of the receptive field. A detailed search of the electron micrography by Dowling and Boycott showed continuous cell-to-cell connections only with amacrines, not with horizontal cells. Thus, this strong evidence indicates that it is the amacrine cells which are responsible for the organization of the periphery of the receptive field. In addition, the amacrines synapse directly with the ganglion cells and the antagonism (or inhibition) of the response of the center of the receptive field could occur directly at the ganglion cell.

On the basis of these deductions a retinal circuit, or wiring diagram, has been constructed by Dowling and Boycott (1966) for a receptive field, Fig. 6.60. In this diagram the ganglion cells connect directly with the bipolars only in the center of the receptive field. The other bipolars in the field connect with the ganglion cell through the amacrine cells. The amacrine cells connect back to the bipolars (p. 218) throughout the field, contact other amacrine cells, and connect with the ganglion in the center of the field. The horizontal cells are also drawn to show back connections with the receptor terminals. For example, if the center of a receptive field is an excitatory "on" response, the signal goes directly from the receptor

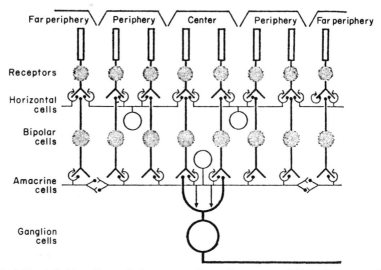

FIG. 6.60   A "wiring diagram" from a ganglion cell receptive field in the mammalian retina. Inverted V's represent excitatory synapses and circular arrows inhibitory synapses. Downward arrows represent dendritic zone. [From Dowling and Boycott (1966).]

through the bipolar to the ganglion. Illumination of the periphery would be inhibitory, and the antagonistic effect would be transmitted through the amacrine cells to the ganglion and result in a weakening of the center-field signal to the ganglion. The role of the horizontal cells is not clear, but since they connect only a few receptors they possibly sum the response of several receptors, thereby creating the center-field response.

PATHWAY FROM THE OPTIC NERVE

In the preceding sections we have seen that most of the information processing of visual stimulation occurs within the retina itself. Over 100 million photosensors which cause a change in potential upon stimulation have their information reduced to about 1 million optic nerves, which propagate this information in the form of spikes, or action potentials, to the brain.

The subsequent path is illustrated in Fig. 6.61. It is seen that the parts of the optic nerves from the nasal half of the retina cross (decussate), while the optic nerves from the temporal (outside) parts of the retina do not. The region of the cross is called the optic *chiasma*. Consider a light stimulus from the left, as indicated by the ray lines at the top of the figure. The stimulus from the left eye crosses over to the right side of the brain, while that of the right eye goes directly to the right side (black lines). Thus, the

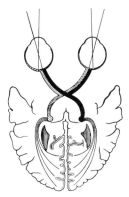

FIG. 6.61   The visual pathway showing partial decussation in the chiasma. Note that all impulses from one visual field pass to the same side of the brain. [From Starling and Evans (1962).]

stimulus from the left proceeds in its entirety to the right side of the brain; similarly a stimulus from the right goes to the left side of the brain. The degree of decussation varies with the degree of binocular vision and with position on the phylogenetic scale. In the opossum 20% of the optic nerves remain uncrossed, in the rat about 10%, in the guinea pig 1%, and in the cat and man, both with frontally directed eyes, approximately equal numbers are crossed and uncrossed.

The signals from the optic nerves synapse at the *lateral geniculate body* (geniculate means knee-shaped) within the brain. From there, new fibers called the *optic radiation* convey the signal to the *visual cortex*, or *striate* area. It is called striate because of its striped appearance.

The size of the striate area varies in humans, the average being about 2600 mm^2. The area of the retina of a normal human adult is about 350 mm^2, so the ratio of area of the striate to the retina is about 7 to 1. There is a precise mapping of the visual field on the striate with appropriate striate area magnification factors which correspond to the varying density of receptors in the retina. In fact, it is customary to speak of a "point-to-point" localization of retinal projection on the cortex. Lesions of the striate area lead to blindness in well-defined areas of the visual field. Holmes (1945), in a systematic study of the effects of gunshot wounds, was able to develop the mapping illustrated in Fig. 6.62. The right side of this figure shows regions of the right half of the visual field, while the corresponding markings on the left side of the figure shows the location on the striate of these regions. Also indicated in this figure is the varying magnification factor mentioned above. While the above ratio of 7 to 1 was the ratio of areas, the usual terminology is millimeters of cortex to degrees of visual field; 1 mm of the retina corresponds to about 3.3° of visual field. More

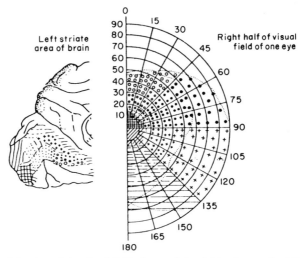

FIG. 6.62   Mapping of the visual field of part of the right retina on the left striate cortex. [From Holmes (1945).]

careful mapping of the striate has shown that the magnification factor in the monkey and baboon decreases evenly from about 5.6 mm/deg at the fovea to about 0.1 mm/deg at 60° from the optic axis.

As might be expected from the above discussion, a localized wound on one side of the striate area in man causes blindness in that part of the visual field projected on that area of the cortex. Damage to the entire striate area on one hemisphere of the brain causes blindness for half the visual field in both eyes. This is called *hemianopia*. In the converse of this, it has been shown that stimulation of an exposed striate causes images of light, shadow, and color. Experiments in which growing monkeys have had vision deprived in one eye have shown abnormalities in the development of the visual cortex (Hubel 1979). It is therefore important that children with vision defects wear corrective glasses at an early age so that the cortex can develop normally even if glasses are not worn regularly at a later age.

THE VISUAL CORTEX

It must be kept in mind throughout the discussions of these experiments that different species are being tested and the necessary information for survival differs. Therefore, the sensitivity to passing shadows or slits of light, their direction, and angle will be weighted differently than in man. In fact, in some species retinal sensitivity to different types of motion has been found to be more sensitive than in primates. The experiments on rabbits, described earlier, yield somewhat different results than those on cats to be described in this section and, of course, future experiments on primates will reveal further differences. However, all the results show what is possible and also confirm that, because of its information processing character, the eye may be considered as an extension of the brain.

Microprobes inserted into cells in the visual cortex of cats yield varieties of discharges for different light intensities and motion on the receptive field in the retina connected to the striate cells. A systematic study of these cells has been made by Hubel and Wiesel (1959, 1961, 1962, 1965). They have categorized visual cortex cells into four basic types: simple, complex, hypercomplex and higher order hypercomplex. Each will be described briefly.

Simple cells. These have a narrow bar of excitatory sensitivity within a receptive field flanked on either side by an antagonistic, or inhibitory field. Some examples are shown in Fig. 6.63, where the × stands for the "on" response and the Δ is the "off" response. The receptive field axes are indicated. An important characteristic of these is that a precise angle of the receptive field is required, not varying from maximum by more than 5–10°. Hundreds of cortical neurons of this type have been observed, with all possible angles being represented.

Complex cells. There are a variety of types of behavior exhibited by complex cells: activated by a slit, by an edge, or by a dark bar. Usually

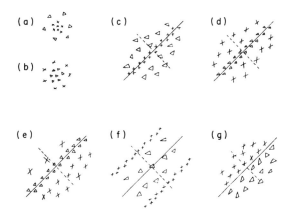

FIG. 6.63 Common arrangements of lateral geniculate and cortical receptive fields. (a) "On"-center geniculate receptive field. (b) "Off"-center geniculate receptive field. (c)–(g) Various arrangements of simple cortical receptive fields. X, areas giving excitatory responses ("on" responses); Δ, areas giving inhibitory responses ("off" responses). Receptive-field axes are shown by continuous lines through field centres; in the figure these are all oblique, but each arrangement occurs in all orientations. [From Hubel and Wiesel (1962).]

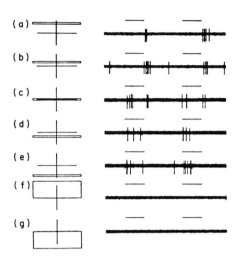

FIG. 6.64 Responses of a cell with a complex receptive field to stimulation of the left (contralateral) eye. Receptive field located in area contralis. The diagrams to the left of each record indicate the position of a horizontal rectangular light stimulus with respect to the receptive field, marked by a cross. In each record the upper line indicates when the stimulus is on. (a)–(e), stimulus $\frac{1}{8} \times 3°$, (f)–(g), stimulus $1\frac{1}{2} \times 3°$, (4°, is equivalent to 1 mm on the cat retina). Cell was activated in the same way from right eye, but less vigorously, Positive deflections upward; duration of each stimulus 1 sec. [From Hubel and Wiesel (1962).]

there is a preferred orientation and maximum response is obtained when the light, or edge, is in motion. The position of the light, indicated on the left by the rectangle representing the slit, with respect to the central field indicated by the crossed horizontal and retinal lines, is shown on the left in Fig. 6.64. The lines above the discharge data indicate the time the light is on; duration 1 sec. The response of (a) and (b) indicates that the region above the horizontal lines is an "off" position, while that for (d) and (e) indicate that the area below the horizontal is an "on" position. (c) is clearly an "on-off" position. Increasing the size of the slit in (f) and (g) gave no response in either region of the receptive field. For this reason, i.e., response only to a slit, complex cells differ from the simple cells.

Hypercomplex cells. As with the other cells, there are a variety of types of hypercomplex cells, but a general characteristic is their response to a line that stops, an edge, or a corner. The behavior of one of these cells is shown in Fig. 6.65. The shaded area to the left represents a shadow edge being moved up or down in the direction of the arrow. The field affected is indicated by the dashed rectangle with a bisector dividing it into two parts, each about 2×2 deg. It is first seen by the discharge that response is obtained only by upward motion. Further, (a), (b), and (c) indicate that the

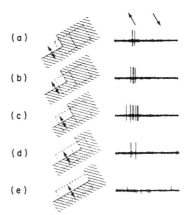

FIG. 6.65   Records from a hypercomplex cell. Stimulation of right eye. Receptive field, 2 × 4 deg, indicated by interrupted rectangle. Stimulus consisted of an edge oriented at 2:00, with dark below, terminated on the right by a second edge intersecting the first at 90 deg. (a)–(c): up-and-down movement across varying amounts of the activating portion of the field; (d)–(e): movement across all of the activating portion and varying amounts of the antagonistic portion. Rate of movement 4 deg/sec. Each sweep 2 sec. [From Hubel and Wiesel (1965).]

response increases the greater the exposure of the left half of the field to the moving edge. (d) and (e) show that as the right half is exposed it inhibits the response. From these and other experiments on the same cell, with different angles and lengths of the shadow, the cell was found to be responsive to a particularly oriented edge, provided the edge was limited in length.

Higher Order Hypercomplex cells. These cells are given a separate designation because their behavior cannot be described in the relatively simple terms of the other cells. One of the main characteristics of higher order cells is an ability to respond to two sets of stimuli with orientations 90 deg apart. An example is shown in Fig. 6.66. With a stimulus stopped to the left, Fig. 6.66b, the cell responded to movement down or up throughout an area indicated by the left dashed rectangle (this area is actually oval in the animal). With the edge stopped to the right, Fig. 6.66d, the cell responded over a separate region indicated by the right dashed rectangle. Corners made from an edge with dark below ((a) and (c)) were almost ineffective. A dark tongue 5° wide made from the combination of the two corners, Fig. 6.66f, evoked a brisk response when moved in from above.

A possible scheme for the organization of simple receptive fields is shown in Fig. 6.67. The ganglion cells of the retina synapse at the lateral geniculate cells in the brain, which produce a similar response to the visual

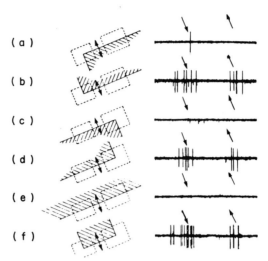

FIG. 6.66   Responses of a higher order hypercomplex cell. Receptive field 6 deg, from center of gaze. Regions from which responses were evoked are indicated roughly by the two interrupted rectangles, of which the right was 3 ×4 deg in size. A right-angled corner with dark up and to the right (b) evoked responses when moved down or up over the left-hand region; no response was evoked in the right-hand region. A corner with darkness up and to the left evoked responses to up-and-down movement only over the right-hand region (d). Little or no response was obtained with corners with dark below [(a) and (c)], or with unstopped edges (e). A tongue (f), combining stimuli (b) and (d), gave the most powerful responses. Rate of movement, about 0.5 deg/sec. Duration of each sweep, 20 sec. [From Hubel and Wiesel (1965).]

cortex. Four of these cells are indicated on the right in Fig. 6.67. Each receives information from receptive fields of the retina whose "on" centers are arranged in a straight line, indicated by the heavy dashed line on the left. The signals from the four lateral geniculate cells all converge on the cell on the lower right. This is the position scheme proposed by Hubel and Wiesel (1962) to explain the behavior of simple cells.

Their scheme for complex cell behavior is shown in Fig. 6.68. The three cells on the right are simple cells which synapse with a complex cell. The receptive fields of the simple cells are shown on the left in the dashed rectangle with their boundaries staggered. Any vertical-edge stimulus falling across this rectangle, regardless of its position, will excite some simple cells and lead to excitation of the complex cell. Figure 6.69 shows their scheme for a hypercomplex cell synapsing with an excitatory and an inhibitory complex cell, each of which responds to its own region of the field shown on the left. In the situation shown, the edge covers both fields equally and the result is no excitation.

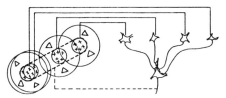

FIG. 6.67 Possible scheme for explaining the organization of simple receptive fields. A large number of lateral geniculate cells, of which four are illustrated in the upper right in the figure, have receptive fields with "on" centers arranged along a straight line on the retina. All of these project upon a single cortical cell, and the synapses are supposed to be excitatory. The receptive field of the cortical cell will then have an elongated "on" center indicated by the interrupted lines in the receptive-field diagram to the left of the figure. [From Hubel and Wiesel (1962).]

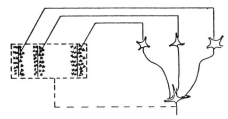

FIG. 6.68 Possible scheme for explaining the organization of complex receptive fields. A number of cells with simple fields, of which three are shown schematically, are imagined to project to a single cortical cell of higher order. Each projecting neuron has a receptive field arranged as shown to the left: an excitatory region to the left and an inhibitory region to the right of a vertical straight-line boundary. The boundaries of the fields are staggered within an area outlined by the interrupted lines. Any vertical-edge stimulus falling across this rectangle, regardless of its position, will excite some simple-field cells leading to excitation of the higher-order cell. [From Hubel and Wiesel (1962).]

FIG. 6.69 Wiring diagrams that might account for the properties of hypercomplex cells. A hypercomplex cell responding to a single stopped edge receives projections from two complex cells, one excitatory to the hypercomplex cell (E), the other inhibitory (I). The excitatory complex cell has its receptive field in the region indicated by the left (continuous) rectangle; the inhibitory cell has its field in the area indicated by the right (interrupted) rectangle. The hypercomplex field thus includes both areas, one being the activating region, the other the antagonistic. Stimulating the left region alone results in excitation of the cell, whereas stimulating both regions together is without effect. [From Hubel and Wiesel (1965).]

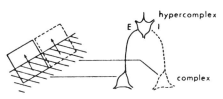

COLOR SPECIFICATION

The science of colorimetry has developed from the necessity of color matching, i.e., given a sample color such as a chip of paint, one wishes to know how to match it. The basic colors one has to use are the visible spectrum, such as that which is seen by refraction through a prism or other type of monochrometer. If one performs an analysis of the reflected light from the paint chip with a spectrophotometer, which measures the relative intensity of each wavelength of light, a broad band will result with a maximum at the dominant color. Thus, given any number of mono-chromators, how should a piece of white paper be simultaneously illumi-nated with different spectral colors with different intensities to match the unknown?

For the purpose of our discussion we will express the visible spectrum in terms of wavelength range. The spectral or rainbow colors are

violet	400–450 nm
blue	450–500
green	500–570
yellow	570–590
orange	590–610
red	610–700

Note that the retina is sensitive to light of shorter wavelength than 400 nm but the lens absorbs it. An aphasic eye, one with the lens removed, can see ultraviolet and it has been reported that even x rays can be seen.

In the matching of a color, which is seen by the reflection from its surface, it is assumed that all spectral colors of the illuminating source have the same intensity. Thus, daylight is the usual standard illuminant but, even with this, something illuminated by the blue sky has different reflection characteristics than when illuminated by the sun. A bright cloudy day will illuminate an object with a reasonably constant intensity at all visible wavelengths. Problems arise in making a match under artificial light and the standardization of illuminants has been developed (see Judd and Wyszecki (1975)].

It has long been known that any color can be matched by mixing the light from three sources of primary colors in the proper proportions. The observer looks into an instrument with a suitable photometric field. The light whose color is to be matched is introduced into one-half of the field and the light from three primary sources illuminate the other half. The observer adjusts the intensities of the three until the color is matched. The most satisfactory instrument for this is called the *flicker photometer*. This is based on the fact that the fusion frequency in the eye of repetitive stimuli is a function of their brightness. In this instrument the two stimuli are alternated at about 15 flashes/sec. If the illuminations of the two surfaces have different intensities, a sensation of flicker is experienced, and adjustments are made in the intensities until the flicker disappears. The unknown color can then be specified by three numbers, X, Y, Z, which represent the amount of each of the three primaries. These numbers are known as the *tristimulus values*. Note here that two different observers will occasionally specify slightly different numbers; this will be discussed later.

This technique of color matching has been used to construct a table of tristimulus values of each of three primaries required to match light of nearly a single wavelength for all visible wavelengths taken at 10-nm intervals. This was done in England independently by Wright (1928) and Guild (1931). Wright used the spectrum colors 460, 530, and 650 nm as primaries while Guild used red, green, and blue filters. Therefore, the tristimulus value tables of each were different. The choice of primaries is not important, however, since they are but intermediates through which a given color can be produced. A simple transformation can convert one system to another, and it was found that their tables were in very good agreement.[†]

The primaries selected by both Wright and Guild are inconvenient to use because both systems require negative quantities to match some colors. A negative quantity is obtained by adding one of the primaries to the opposite side of the field, where it is combined with the color whose tristimulus values are being determined. It can easily be shown (Hardy, 1936) that no set of real primaries exists which does not require both positive and negative values.

In 1931 the ICI (International Commission on Illumination) adopted three artificial primaries, $\bar{x}$, $\bar{y}$, and $\bar{z}$; artificial in that they do not exist as real colors. The wavelengths of each of the spectral colors is then expressed as quantitites of each of these primaries. A plot of this table is shown in Fig. 6.70. Any wavelength on the abscissa can be produced by the sum of the amounts of $\bar{x}$, $\bar{y}$, $\bar{z}$ at that wavelength. Although artificial, these primaries have the general colors of $\bar{x}$ (reddish–purple), $\bar{y}$ (green), and $\bar{z}$ (blue). When matching spectral wavelengths $\bar{x}$, $\bar{y}$, $\bar{z}$ are used, but in matching colors not of a single wavelength the tristimulus values X, Y, Z are used. The shape of the curve of Y was chosen to match the photopic sensitivity curve of the eye (Fig. 6.12). (It should be noted that later studies, Fig. 6.14, show this to be not quite correct.) With this choice, the

---

[†] This transformation has the following form. $r' = k_1 r + k_2 g + k_3 b$, $g' = k_4 r + k_5 g + k_6 b$, $b' = k_7 r + k_8 g + k_9 b$, where r, g, and b are the tristimulus values of one set of primaries, r', g', and b' are the tristimulus values of the second set, and $k_1 \cdots k_9$ are the tristimulus values of the first set of primaries on the basis of the second set. It is seen from this that the transformation is of a simple matrix form [see Judd and Wyszecki (1975)].

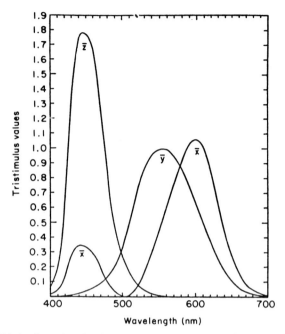

FIG. 6.70   Tristimulus values for the various spectrum colors. The values of $\bar{x}$, $\bar{y}$, $\bar{z}$ are the amounts of the three I.C.I. primaries required to color match a unit amount of energy having the indicated wavelength. [Reprinted from A. C. Hardy, "Handbook of Colorimetry" by permission of the MIT Press. Copyright 1936 by the Massachusetts Institute of Technology.]

brightness of a color can be specified by the Y value. For example, if two colors are matched by $X = 15$, $Y = 25$, $Z = 6$, and $X' = 30$, $Y' = 50$, $Z' = 12$, one can immediately state that they are the same color with one being twice as bright as the other, since $Y' = 2Y$. Furthermore, the Y value scale is between 0 (black) and 100 (white). Thus, the above color with $Y' = 50$ has brightness 50% of that of a white standard illumination.

The evaluation of the quality of a color (chromaticity) is obtained by defining three tristimulus coefficients, x, y and z, as

$$x = \frac{X}{X + Y + Z}, \qquad y = \frac{Y}{X + Y + Z}, \qquad z = \frac{Z}{X + Y + Z} \qquad (6.23)$$

and it is seen that the sum of these is unity

$$x + y + z = 1 \qquad (6.24)$$

With this relation only two coefficients, usually x and y, are required because z can then be determined from Eq. (6.24). It should now be evident that instead of using the tristimulus values X, Y, Z, even more information can be supplied if a color is specified in terms of Y, x, and y.

As an example, (Hardy, 1936) suppose two colors have tristimulus values $X = 15.50$, $Y = 24.19$, $Z = 22.64$, and $X' = 25.26$, $Y' = 39.42$, $Z' = 36.89$. In the new notation these values become, respectively, $Y = 24.19$, $x = 0.2487$, $y = 0.3881$ and $Y' = 39.42$, $x' = 0.2487$, $y' = 0.3881$. It is immediately evident that the two colors have the same chromaticity with only a difference in brightness.

A convenient way to examine color is the chromaticity diagram, in which the x and y chromaticity coefficients are the abscissa and ordinate of a cartesian system, respectively. This is shown in Fig. 6.71, where the solid curve represents the locus of pure spectral colors computed from the data of Fig. 6.70, using Eqs. (6.23). Note that from Eq. (6.24) only x and y are required and, for example, white light of value $x = y = z = 0.333$ can be plotted with only the two coordinate points, $x = 0.333 = y$. This is marked W on Fig. 6.71. Points G and R represent a certain green and red, respectively. When these are additively mixed, any proportion of them must lie on the line connecting them. This should be evident by the algebra

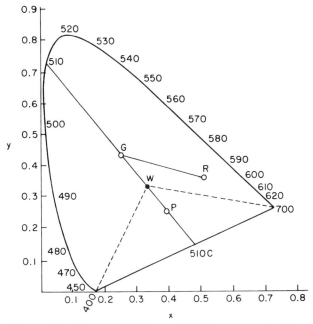

FIG. 6.71 Chromaticity diagram with locus of visible pure wavelengths. W, perfectly balanced illuminant of white light. Colors in region below dashed lines cannot be formed by simple addition of two pure wavelengths. G and R, two typical green and red colors. All colors on a line connecting these can be formed by simple addition of various amounts of G and R.

of adding coefficients, and is done in detail by Hardy (1936). Because of this additivity relation, it is clear that the lower solid line connecting 400 nm with 700 nm of the locus of spectral colors forms the boundary of all real colors, since all real colors must be considered as the sum of spectral components in various proportions.

By logical extension of this thought, it is seen that all real colors above the dashed line in Fig. 6.71 can be represented by the sum of white light W and light of a certain wavelength. In this figure an arbitrary green, marked G, is connected by a line from W to spectral wavelength 510 nm. This wavelength is known as the dominant wavelength and the *purity* of the color is specified by the fraction of the total length of the line that lies between W and G. In this diagram the spectral color of the dominant wavelength is 510 nm and the purity of G is about 25%. If G lay on the locus of spectral colors, it would have a purity of 100% and the color would be called *saturated*, i.e., no white. Note here that the three primaries of Fig. 6.70 selected by the ICI have a higher saturation than any spectral color. This was done to avoid the use of negative values and it is for this reason that the primaries are called artificial.

When a line joining two colors passes through the illuminant point, in this case W, they are said to be complimentary colors and, when they are added in appropriate proportions, they will produce luminant W. If the line is extended in both directions to the locus of spectral colors then these dominant colors are complimentary. It is seen in Fig. 6.71 that the dashed lines represent the two limits of spectral complimentarity, that is, no spectral colors lie on the straight line which encloses the diagram at the bottom. It should be noted that above the dashed lines only two spectral colors need be added to produce white. Below the dashed lines is the region of purples and magentas. These clearly cannot be obtained by mixing white light with any single pure spectral color. An artifice is used to specify the purples. The compliment of a color always lies on the opposite side of the illuminant point. Thus, in Fig. 6.71 the specification of the purple is obtained by extending the line through P and W in both directions. The intersection on the base line is labeled, in this case, 510c, which means that it is the compliment of 510-nm green, and its purity is its fraction between W and the intersection with the closure bottom line, in this case about 40%. In general, the colors of this chromaticity map are shown in Fig. 6.72, and it is seen from this discussion that the specification by brightness, dominant wavelength, and purity correspond to the psychological attributes brilliance, hue, and purity (percent saturation), respectively.

It should be noted that the shape of the chromaticity diagram is the result of the shapes of the three human color absorption curves. In the reproduction of all colors three sources of comparable wavelength–

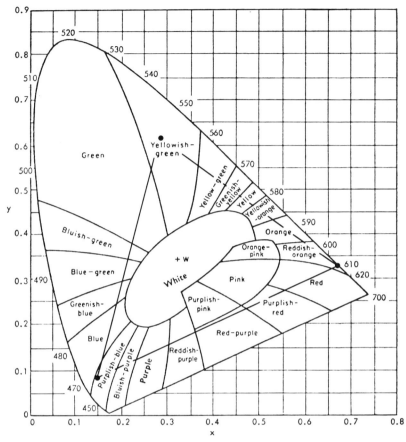

FIG. 6.72 Common names of color regions on chromaticity diagram. Three points are colors of typical color TV phosphors and region within the connecting lines are those colors that can be obtained by mixing of outputs of these three phosphors. [From Kelly (1943).]

brightness distribution are required. These do not exist. In color printing the most inexpensive way is to use small dots, below the resolution of the eye, of three primary colors. If their reflection chomaticities are plotted on Fig. 6.72, the possible color combinations would lie within the triangle formed by the lines connecting the points (if printed on white paper). More expensive color printing processes may use up to 20 different colored inks to match more closely the visual enclosure formed by the locus of pure spectral colors. Thus, these colors plus white may well reproduce most colors within human color discrimination.

The introduction of color television presented difficult problems. In color television each picture point of the screen has three small dots of

TABLE 6.5[a]

Color	Compound	Chromaticity coefficients	
red	europium yttrium vanadate	$x_R = 0.68$	$y_R = 0.32$
green	zinc cadmium sulfide	$x_G - 0.28$	$y_G = 0.60$
blue	zinc sulfide	$x_B = 0.15$	$y_B = 0.07$

[a] From Judd and Wysecki (1975).

phosphors which, upon being struck with electrons, luminesce in their characteristic colors proportionately to the number of electrons striking them. The twofold problem was to find phosphors which have both the appropriate colors and comparable brightness. Compromises had to be made because of conflicting desires to have maximally colored pictures with maximum brightness. The most commonly used phosphors with their chromaticities are shown in Table 6.5. These colors are plotted as the solid points connected by straight lines in Fig. 6.72. As discussed earlier, only those colors which lie on or within the connecting lines can be produced. It is seen that much of the purple field as well as the blue-greens have been sacrificed and that the best red has a somewhat orange-red appearance.

COLOR VISION

The understanding of the mechanism of color vision in humans is still under investigation and the literature of models and criticism is extensive. Because of this, only a brief survey will be given, and the interested reader might start with the book by Sheppard (1968) who has reduced the important references at the date of writing to 245.

The discussion in the previous section of chromaticity indicated that three primaries are all that are needed to specify the characteristics of any color. With the recognition that nature seeks simplicity, a search for three color receptors began. As shown earlier in this chapter, rods are responsible for scotopic vision and are not present in the fovea. Furthermore, light of sufficient intensity for photopic vision bleaches rhodopsin. Only cones are in the fovea and these are known to be responsible for color vision. Rhodopsin has been extracted from the retina, and the spectroscopic absorption spectrum, as well as bleaching characteristics, have been shown to correspond to those in the living eye. A chemical model of the formation of other retinal opsins which can have different spectral absorption characteristics was shown in Table 6.2. In primates, despite many attempts, the *only* visual photopigment which has been obtained by extraction is rhodopsin, and therefore some investigators seriously question even the existence of other photopigments. However, as we shall see, the evidence that there are three types of photopigments in the cones is fairly persuasive, although their nature turns out to be somewhat unexpected.

Early work on color blindness strongly indicated the presence, or absence, of three color receptors. Detailed studies on many individuals showed that human color vision can be divided into three main classes (these classes have subdivisions which will not be considered here). The three classes are (Pitt, 1944):

Trichromats. The members of this class have color vision that has three variables and may express any color C by the tristimulus values of three primaries,

$$C = X + Y + Z$$

where, as discussed in the preceding section, XYZ are generally red, green, and blue, respectively. Note that this class can be subdivided into those that have normal color vision and those with abnormal color vision, called anomalous trichromats.

Dichromats. The members of this class have color vision which is a function of two variables and are subdivided into (a) *protanopes*, called red-blind, (b) *deuteranopes*, called green-blind, and (3) *tritanopes*, called blue-blind.

Monochromats. The members of this class have color vision, which is a function of a single variable.

Tritanopes are fairly rare and, as will be seen, blue-absorbing cones are difficult to measure. Therefore, the most meaningful measurements on the color absorption of red and green cones are on protanopes and deuteranopes. Pitt confirmed that protanopia is caused by the absence of red sensation and that deuteranopia is caused by the red and green sensations being identical. Figure 6.73 shows the luminosity curves for protanopes and deuteranopes.

Rushton, who studied the bleaching and recovery of rhodopsin by reflected light (Fig. 6.13), performed similar experiments on color-blind subjects. The protanope presumably has but one pigment in the red-green range. The experiments of Rushton (1963) showed that, like rhodopsin, this pigment called *chlorolabe* (green taking) bleaches and recovers. This bleaching with increasing light intensity is shown in Fig. 6.74, and the intensity notation is in log trolands (e.g., 4.5 log td = $10^{4.5}$ td). The bleaching is quite rapid, compared to rhodopsin, as is the recovery, shown by the solid curve. This experiment clearly shows the existence of a pigment with the peak of its absorption spectrum at about 540 nm. Similar experiments were performed on the deuteranope who is green-blind. The peak of his

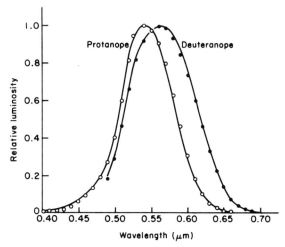

FIG. 6.73   Deuteranopic and protanopic luminosity curves. [From Pitt (1944).]

single photoabsorption curve is at about 560 nm. The pigment of the deuteranope in the red-green region is called erythrolabe (red taking). Presumably there is another pigment in the blue region, called *cyanolabe*, but it has not been detectable with the densitometer.

An indirect measurement of three photopigments in human retina has been made by Brown and Wald (1964). A portion of human retina shortly

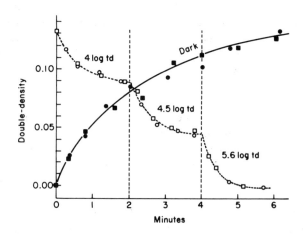

FIG. 6.74   Double-density of protanope pigment measured during 2 min bleaching in succession by lights of retinal illuminations shown ($4.5 \log td = 10^{4.5}td$). Black symbols represent subsequent regeneration in the dark. Circles and squares are repetitions of the same experiment. [From Rushton (1963).]

after death of the donor was studied by absorption difference spectra. A spectrophometer trace of single cones was made before and after bleaching, and the difference was assigned to the absorption characteristics of the three photopigments found. These are shown in Fig. 6.75, and they would serve as suitable primaries in a chromaticity diagram; the blue has a peak between 440 and 450 nm, green between 520 and 540 nm, and "red" between 550 and 580 nm. Note that the "red" actually has its peak in the yellow-green region of the spectrum. The most severe criticism of the conclusion from these measurements is that they do not take into consideration the alterations which may arise from waveguide physics. The diameter of the cones is about 1–1.5 μm, which is the order of 1000 nm. They are therefore only about twice the wavelength of the incident light. Enoch (1961) has showed in the human retina that modal transmission is characteristic of that of waveguides (see also Snitzer and Osterberg (1961)). This changes the transmittance wavelength relationship and introduces a greater effect of transmittance with obliqueness of the radiation. Thus, although the peaks of Fig. 6.75 may be in correct positions, the distribution of the shape is not guaranteed unless it is known positively that there was no obliqueness of the incident illumination. Wald (1964) subsequently corrected the magnitude of the peaks of Fig. 6.75 for the absorption of the nonvisual pigments of the eye and the relative numbers of the cones containing the different pigments, Fig. 6.76. These curves correlate with psychophysical measurements of relative luminosities of the colors. A more recent measurement, with results similar to Wald's, has been made by Bowmaker and Dartnall (1980).

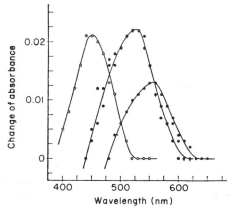

FIG. 6.75 Difference spectra of the visual pigments in single cones of the human parafovea. In each case the absorption spectrum was recorded in the dark, then again after bleaching with a flash of yellow light. They involve one blue sensitive cone, two green-sensitive cones, and one red-sensitive cone. [From P. K. Brown and G. Wald, *Science* **144**, 45 (1964). Copyright 1964 by the American Association for the Advancement of Science.]

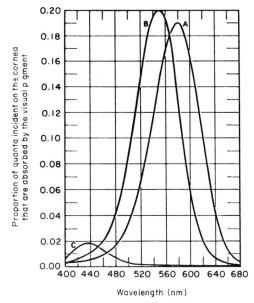

FIG. 6.76 Estimates of the absorption spectra of the three-color-system in a human subject with normal color vision. These curves are actually derived from psychophysical measurements that agree well with direct physical measurement. They are estimates of the absorption spectra of the visual pigments as seen from the cornea, that is, after having been affected by the absorption of the nonvisual pigments of the eye, and by the relative numerosities of cones containing the different pigments. The C curve is about three times higher in this subject than that in the average subject. [From G. Wald, *Science* **145**, 1007 (1964). Copyright 1964 by the American Association for the Advancement of Science.]

Even without knowing the actual mechanism, certain characteristics of human color vision have been determined. In the preceding section, color differences of different illuminants were shown. The eye is capable of making correct color matches. When the eye adapts to colored illumination, it apparently does so through a change in the relative sensitivity of the color receptors. When adaptation has taken place, if there are no sudden abrupt transitions at some wavelength, the new illuminant color appears white. This is illustrated in Figs. 6.77a and 6.77b. Suppose the eye is viewing a color of wavelength M in Fig. 6.77a under conditions of daylight illumination. M will stimulate certain fractions of the red and green color receptors. In Fig. 6.77b the eye views the same color under incandescent light which has an excess of orange. The red and green receptors reduce their sensitivity and a color match of M can still be made, but with a different ratio of receptor stimulation.

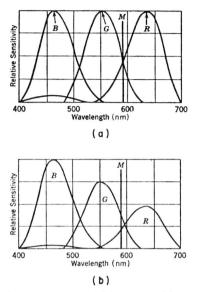

FIG. 6.77   (a) Schematic eye-receptor sensitivity curves with a monochromatic line stimulating the red and green receptors equally and causing a yellow sensation. (b) Schematic eye-receptor sensitivity curves after the eye has been adapted to an orange source. The red receptor is now less sensitive with respect to the green receptor than before, and the monochromatic line will appear slightly more green than before adaptation took place. [From Evans (1948).]

The sensitivity of the eye to incremental color shifts has also been determined and, as expected, it is not constant. Figure 6.78 illustrates the locus of just noticeable differences in color on the chromaticity diagram. The axes of each ellipse have been multiplied by ten for clarity. In hearing, each fractional note activates a different nerve (or brain) response so that the hearing of notes is distinct. For instance, not only can a G chord be distinguished from a pure G note, but a trained musician has no trouble identifying the individual notes in a chord no matter how complex. In contrast to hearing, what we must conclude in this section is that all humans are color-blind. This blindness is demonstrated in color matching. When a pure color is matched by the addition of three primaries, we cannot tell the difference nor can we distinguish the three primaries.

Colored light must pass through the lens of the eye, which yellows with age, as well as the *macular*, or yellow, pigment in front of the fovea. These two substances cause selective absorption of the spectrum and transmission density versus wavelength for a 21-yr-old human eye is shown in Fig. 6.79. The lens pigment prevents the retina from being overstimulated or

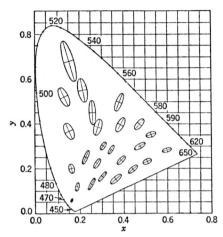

FIG. 6.78   Ellipses indicate just detectable changes of chromaticity by the human eye. The axes of each ellipse have been multiplied by 10. [From MacAdam (1942).]

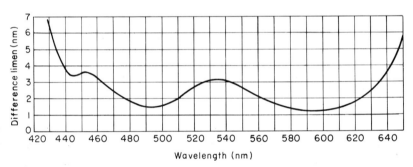

FIG. 6.79   Hue discrimination curve for normal human. [From Wright and Pitt (1934).]

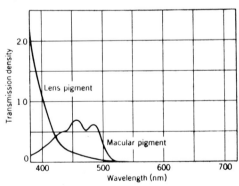

FIG. 6.80   Spectral transmission density of the lens pigment and the macular pigment of the average 21-yr-old human eye. [From Judd and Wyszecki (1975).]

damaged by ultraviolet light, while the macular pigment apparently exists to prevent overstimulation by blue light, which causes long-lasting after-images. Because of this understimulation of the blue cones, they have a greater sensitivity for blue so that colors appear normal. However, in measuring the blue, or cyanolabe, pigment the macular pigment is deletrious and has reduced the reliability of data, Fig. 6.80. It also affects certain kinds of color measurement. The macular pigment can be noticed if one closes his eyes for about 20 sec, then opens them and stares at a sheet of white paper. A yellow spot will appear for a few seconds before the cones have had time to develop all of their blue supersensitivity relative to the other cones.

Just as skin, hair, and iris pigmentations differ with individuals, so apparently does macular pigment. One would then not expect all individuals to obtain identical color matchings. Furthermore, there is a variation among individuals in either the sensitivity or the quantity of chlorolabe and erythrolabe which results in a variety of red and green matchings. Figure 6.81 shows luminosity curves of photopic vision for six different persons, with the dashed line being the ICI average. It is seen in this curve, for example, that at 600 nm the red sensitivity can differ by a factor of two. Measurements of partial color blindness were made in two studies on 23,000 school children (see Judd and Wysecki (1975)). Curiously, it was found that while 8% of boys had a color-vision defect, only 0.5% of the girls had such a defect. To account for enough mothers for this 8% of red-green defect boys, about 16% of all mothers had to be genetic carriers of the defect. For a discussion of other color-vision defects see Wald and Brown (1965).

Further experiments have shown that there are probably three color signals which reach the brain from the cones. These colors are nominally

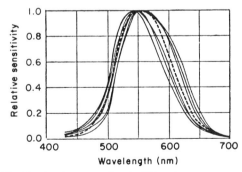

FIG. 6.81   Individual luminosity curves for six different persons compared to the standard ICI curve and showing the large deviations from person to person. [From Gibson and Tyndall (1925).]

called R, G, and B, although their maximum sensitivities are at wavelengths of 570, 540, and 450 nm. Note the close correspondence with the Brown and Wald data of Fig. 6.75. Gouras (1968, 1970, 1971) has made microelectrode recordings in ganglion cells of monkeys and has found three distinct groups of cells based on red, green, or blue sensitivity. The work of Wiesel and Hubel (1966) and DeValois (1965, 1971) has shown that the signal reaches the lateral geniculate body in four types of spectrally opponent cells: red-excitatory, green-inhibitory or $(+R - G)$; yellow-excitatory, blue-inhibitory or $(+Y - B)$; and their respective mirror images $(+G - R)$ and $(+B - Y)$. These are comparable in behavior to white–black cells discussed earlier in this chapter, which fire best with white light and are inhibited by the absence of light and their mirror image black – white which fires best with no light and are inhibited by the presence of white light. A $(+R - G)$ will fire in the presence of red light and be inhibited by green light. Gouras (1971) has investigated color sensitive cells in the striate cortex of the monkey and found that they also reflect the excitatory–inhibitory behavior of the lateral geniculate nucleus. However, he has observed other types of cell response to color at the retina and the situation at the present time is not clear.

THE COLOR CHROMOPHORES

In this chapter we have discussed in some detail the chromophore rhodopsin. We have also seen that human color vision apparently arises from the absorption of three additional chromophores, but that detailed chemical studies have failed to isolate them from the retina. This has been a central problem in vision research for many years. It has been recently shown, however, that the color chromophores all arise from the retinal molecule in combination with minor variations of the opsin.

In conventional organic chemistry the absorption spectrum of a molecule is dependent upon the degree of conjugation, or spreading of electrons, along the molecule. It should be recalled from p. 173 that the retinal molecule is bound to the opsin molecule by a protonated Schiff base (SBH^+). The idea that electrostatic interaction between the chromophore and charged or dipolar groups on the opsin may shift the absorption spectrum led Honig et al. (1979) to perform computer calculations of this effect. They positioned external charges and found that there was very little flexibility in the choice of position. A negative charge causes a small blue shift and a larger red shift, respectively, when it is located about 3 Å from the $C = N^+$ bond and $C = C$ bond. This is illustrated schematically in Fig. 6.82 for the calculation done for bovine rhodopsin. The shaded

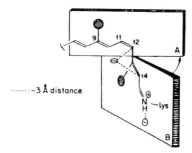

FIG. 6.82   A model for electrostatic interactions in the binding site of bovine rhodopsin. The existence of a counter ion near the protonated nitrogen is assumed. A second negative charge is located ~3 Å above carbon 12. This charge is presumably a member of a charge pair in a salt bridge or possibly the negative end of a neutral dipolar group. [Reprinted with permission from B. Honig *et al.*, *J. Am. Chem. Soc.* **101**, 7084 (1979). Copyright 1979 American Chemical Society.]

circles are CH_3 groups. The proton of the SBH^+ is shown as a $+$, and a counter ion $-$ is assumed. The lysine group of the opsin is indicated by lys. Careful chemistry of the bonding of retinal fragments to various appropriately charged simple groups indicate that this is indeed the probable mechanism, Arnaboldi *et al.* (1979) and Sheves *et al.* (1979).

This is a satisfying conclusion to the problem because it shows that the work of earlier investigators was correct: retinal is the only chromophore-forming molecule that can be extracted from the retina and yet there are chromophores of other colors within the retina.

REFERENCES

Abrahamson, E. W. (1975). Dynamic processes in vertebrate rod visual pigments and their membranes. *Acc. Chem. Res.* **8**, 101.

Arden, G. B., and Low, J. C., (1978). Changes in pigeon cone photocurrent caused by reduction in extracellular calcium activity. *J. Physiol. (London)* **280**, 55.

Arnaboldi, M., Motto, M. G., Tsujimto, K., Balogh-Nair, V., and Nakanishi, K. (1979). Hydroretinals and hydrorhodopsins. *J. Am. Chem. Soc.* **101**, 7082.

Barlow, H. B. (1957); Increment thresholds at low intensities considered as signal/noise discriminations. *J. Physiol. (London)* **136**, 469.

Barlow, H. B. (1958). Temporal and spatial summation in human vision at different background intensities. *J. Physiol. (London)* **141**, 337.

Barlow, H. B., and Fatt, P. (eds.) (1977). "Vertebrate Photoreception." Academic Press, New York.

Barlow, H. B., and Levick, W. R. (1965). Mechanism of directionally selective units in rabbit's retina. *J. Physiol. (London)* **178**, 477.

Bennett, A. G., and Francis, J. L. (1962). Visual optics. *In* "The Eye" (H. Davson, ed.), Vol. 4. Academic Press, New York.

Bertrand, D., Fuortes, M. G. F., and Pochobradsky, J. V. (1978). Actions of EGTA and high calcium on the cones in turtle retina. *J. Physiol. (London)* **275**, 419.

Borsellino, A. and Fuortes, M. G. F. (1968). Responses to single photons in visual cells of *Limulus. J. Physiol. (London)*, **196**, 507.

Bowmaker, J. K., and Dartnall, H. J. A. (1980). Visual pigments of rods and cones in a human retina. *J. Physiol. (London)* **298**, 501.

Brown, P. K., and Wald, G. (1964). Visual pigments in single rods and cones of human retina. *Science* **144**, 45.

Busch, G. E., Applebury, M. L., Lamola, A. A., and Rentzepis, P. M. (1972). Formation and decay of prelumirhodopsin at room temperatures. *Proc. Nat. Acad. Sci. U.S.* **69**, 2802.

Campbell, F. W., and Rushton, W. A. H. (1955). Measurement of the scotopic pigment in the living human eye. *J. Physiol. (London)* **130**, 131.

Cleland, B. G., Dubin, M. W., and Levick, W. R. (1971). Sustained and transient neurones in the cat's retina and lateral geniculate nucleus. *J. Physiol. (London)* **217**, 473.

Cornsweet, T. S. (1970). "Visual Perception." Academic Press, New York.

Crawford, B. H. (1949). The scotopic visibility function. *Proc. Phys. Soc. London* **62B**, 321.

Crescitelli, F., and Dartnall, H. J. A. (1953). Human visual purple. *Nature (London)* **172**, 195.

Dartnall, H. J. A. (1957). "The Visual Pigments." Methuen, London.

Davson, H., and Eggleton, M. G. (eds.) (1962). "Principles of Human Physiology," 13th ed. Lea and Febiger, Philadelphia, Pennsylvania.

DeValois, R. L. (1965). Analysis and coding of color vision in the primate visual system. *Symp. Quant. Biol.* **30**, 567.

DeValois, R. L. (1971). Contribution of different lateral geniculate cell types to visual behavior. *Vision Res. Suppl.* No. 3, 383.

de Vries, H. I. (1943). The quantum character of light and its bearing upon the threshold of vision, the differential sensitivity and acuity of the eye. *Physica* **10**, 553.

Dowling, J. E. (1965). Foveal receptors of the monkey retina, fine structure. *Science* **147**, 57.

Dowling, J. E. (1970). Organization of the vertebrate retina. *Invest. Ophthal.* **9**, 655.

Dowling, J. E., and Boycott, B. B. (1966). Organization of the primate retina: electron microscopy. *Proc. R. Soc. London Ser. B* **166**, 80.

Dowling, J. E., and Wald, G. (1958). Nutritional night blindness. *Ann. N. Y. Acad. Sci.* **74**, 256.

Enoch, J. M. (1961). Nature of the transmission of energy in retinal receptors. *J. Opt. Soc. Am.* **51** 1122.

Enroth-Cugell, C., and Robson, J. G. (1966). The contrast sensitivity of retinal ganglion cells of the cat. *J. Physiol. (London)* **187**, 517.

Evans, R. M. (1948). "An Introduction to Color." Wiley, New York.

Flaming, D. G., and Brown, K. T. (1979). Effects of calcium on the intensity-response curve of toad rods. *Nature (London)* **278**, 852.

Fuortes, M. G. F. (1959). Initiation of impulses in visual cells of *Limulus. J. Physiol. (London)* **148**, 14.

Fuortes, M. G. F., and Hodgkin, A. L. (1964). Changes in time scale and sensitivity in the ommatidia of *Limulus. J. Physiol. London* **172**, 239.

Gibson, K. S., and Tyndall, E. P. T. (1925). Visibility of radiant energy. *Sci. Papers Nat. Bur. Std.* **19**, 157.

Gouras, P. (1968). Identification of cone mechanisms in monkey ganglion cells. *J. Physiol. (London)* **199**, 533.

Gouras, P. (1970). Trichromatic mechanisms in single cortical neurons. *Science* **168**, 489.

Gouras, P. (1971). The function of the midget cell system in primate color vision. *Vision Res. Suppl.* No. 3, 397.

Graham, C. H., and Hartline, H. K. (1935). The response of single visual sense cells to lights of different wavelengths. *J. Gen. Physiol.* **18**, 917.

Graham, C. H., and Margaria, R. (1935). Area and the intensity-time relation on the peripheral retina. *Am. J. Physiol.* **113**, 299.

Graymore, C. (1970). "Biochemistry of the Eye." Academic Press, New York.

Guild, J. (1931). The colorimetric properties of the spectrum. *Phil. Trans. Roy. Soc. London.* **A230**, 149.

Hagins, W. A. (1972). The visual process: excitatory mechanisms in the primary receptor cells. *Ann. Rev. Biophys. Bioeng.* **1**, 131.

Hardy, A. C. (1936). "Handbook of Colorimetry." The Technology Press, Massachusetts Institute of Technology, Cambridge, Massachusetts.

Hartline, H. K. (1934). Intensity and duration in the excitation of single photoreceptor units. *J. Cell Comp. Physiol.* **5**, 229.

Hartline, H. K. (1940). The receptive fields of optic nerve fibers. *Am. J. Physiol.* **130**, 690.

Hartline, H. K. (1949). Inhibition of activity of visual receptors by illuminating nearby retinal areas in the *Limulus* eye. *Fed. Proc.* **8**, 69.

Hartline, H. K., and McDonald, P. R., (1947). Light and dark adaptation of single photoreceptor elements in the eye of *Limulus*. *J. Cell. Comp. Physiol.* **30**, 225.

Hartline, H. K., and Ratliff, F. (1957). Inhibitory interaction of receptor units in the eye of *Limulus*. *J. Gen. Physiol.* **40**, 357.

Hartline, H. K., Wagner, H. G., and MacNichol, E. F., Jr. (1952). The peripheral origin of nervous activity in the visual system. *Cold Spring Harbor Symp. Quant. Biol.* **17**, 125.

Hartline, H. K., Wagner, H. G., and Ratliff, F. (1956). Inhibition in the eye of *Limulus*. *J. Gen. Physiol.* **39**, 651.

Hecht, S. (1947). Colorblind vision I: Luminosity losses in the spectrum for dichromats. *J. Gen. Physiol.* **31**, 141.

Hecht, S., and Williams, R. E. (1922). The visibility of monocharomatic radiation and the absorption spectrum of visual purple. *J. Gen. Physiol.* **5**, 1.

Hecht, S., Shlaer, S., and Pirenne, H. (1942). Energy, quanta and vision. *J. Gen. Physiol.* **25**, 819.

Hinshelwood, C. N. (1945). "The Kinetics of Chemical Change." Oxford Univ. Press, London and New York.

Holmes, G. (1945). The organization of the visual cortex in man. *Proc. R. Soc. London Ser. B* **132**, 348.

Honig, B., Warshel, A., and Karplus, M. (1975). Theoretical studies of the visual chromophore, *Acc. Chem. Res.* **8**, 92.

Honig, B. *et al.* (1979). An external point-charge model for wavelength regulation in visual pigments. *J. Am. Chem. Soc.* **101**, 7084.

Hubbard, R. (1966). The steriochemistry of 11-*cis*-retinal. *J. Biol. Chem.* **241**, 1814.

Hubbard, R., and Kropf, A. (1959). Molecular aspects of visual excitation. *Ann. N. Y. Acad. Sci.* **81**, 388.

Hubbell, W. L. (1975). Characterization of rhodopsin in synthetic systems. *Acc. Chem. Res.* **8**, 85.

Hubel, D. H. (1979). The visual cortex of normal and deprived monkeys. *Am. Sci.* **67**, 532.

Hubel, D. H., and Wiesel, T. N. (1959). Receptive fields of single neurones in the cat's striate cortex. *J. Physiol. (London)* **148**, 574.

Hubel, D. H., and Wiesel, T. N. (1961). Integrative action in the cat's lateral geniculate body. *J. Physiol. (London)* **155**, 385.

Hubel, D. H., and Wiesel, T. N. (1962). Receptive fields, binocular interaction and functional architecture in the cat's visual cortex. *J. Physiol. (London)* **160**, 106.

Hubel, D. H., and Wiesel, T. N. (1965). Receptive fields and functional architecture in two nonstriate visual areas (18 and 19) of the cat. *J. Neurophysiol.* **28**, 229.

Judd, D. B., and Wyszecki, G. (1975). "Color in Business, Science and Industry," 3rd ed. Wiley, New York.

Kelly, K. L. (1943). Color designations for lights. *J. Res. Nat. Bur. Std.* **31**, 274.

Korenbrot, J. I., Brown, D. T., and Cone, R. A. (1973). Membrane characteristics and osmotic behavior of isolated rod outer segments. *J. Cell. Biol.* **56**, 389.

Kuffler, S. W. (1953). Discharge patterns and functional organization of mammalian retina. *J. Neurophysiol.* **16**, 37.

Kuffler, S. W., and Nicholls, J. G. (1976). "From Neuron to Brain." Sinauer Assoc., Sunderland, Massachusetts.

Mason, W. T., and Lee, Y. F. (1973). Resealing properties of biological membranes, *Nature (London) New Biol.* **244**, 143.

Mason, W. T., Fager, R. S., and Abrahamson, E. W. (1974). Ion fluxes in disk membranes of retinal rod outer segments. *Nature (London) New Biol.* **247**, 562.

Matsumoto, H., and Yoshizawa, J. T. (1975). Existence of a β-ionone ring-binding site in the rhodopsin molecule, *Nature (London)* **258**, 523.

Pauling, L. (1949). Zur *cis-trans*-Isomerisierung von Carotinoiden. *Helv. Chim. Acta* **32**, 2241.

Pirenne, M. H. (1967). "Vision and the Eye," 2nd ed. Associated Book Publ., London.

Pitt, F. A. G. (1944). The nature of normal trichromatic and dichromatic vision. *Proc. R. Soc. London Ser. B* **132**, 101.

Polyak, S. (1941). "The Retina." Univ. of Chicago Press, Chicago, Illinois.

Polyak, S. (1957). "The Vertebrate Visual System." Univ. of Chicago Press, Chicago, Illinois.

Ratliff, F. (1965). "Mach Bands." Holden-Day, San Francisco, California.

Ratliff, F., and Hartline, H. K. (1974). "Studies on Excitation and Inhibition in the Retina." Rockefeller Univ. Press, New York.

Ratliff, F., Hartline, H. K., and Miller, W. H. (1963). Spatial and temporal aspects of retinal inhibitory interaction. *J. Opt. Soc. Am.* **53**, 110.

Rentzepis, P. M. (1978). Picosecond chemical and biological events. *Science*, **202**, 174.

Rose, A. (1942). The relative sensitivities of television pick-up tubes, photographic film, and the human eye. *Proc. Inst. Radio. Eng.* **30**, 293.

Rose, A. (1948). On the sensitivity performance of the human eye on an absolute scale. *J. Opt. Soc. Am.* **38**, 196.

Rose, A. (1973). "Vision, Human and Electronic." Plenum Press, New York.

Rushton, W. A. H. (1959). A theoretical treatment of Fuortes' observations upon eccentric cell activity in *Limulus*. *J. Physiol (London)* **148**, 29.

Rushton, W. A. H. (1961). Peripheral coding in the nervous system. *In* "Sensory Communication" (W. A. Rosenblith, ed.). MIT Press, Cambridge, Massachusetts.

Rushton, W. A. H. (1963). A cone pigment in the protanope. *J. Physiol. (London)* **168**, 345.

Sheppard, J. J. (1968). "Human Color Perception." American Elsevier, New York.

Sheves, M., Nakanishi, K., and Honig, B. (1979). Through-space electrostatic effects in electronic spectra. Experimental evidence for the external point-charge model of visual pigments. *J. Am. Chem. Soc.* **101**, 7086.

Snitzer, E., and Osterberg, H. (1961). Observed dielectric waveguide modes in the visible spectrum. *J. Opt. Soc. Am.* **51**, 499.

Starling, E., and Evans, L. (1962). "Principles of Human Psyiology" (H. Davson and M. G. Eggleton, eds.), 13th ed. Lea and Febiger, Philadelphia, Pennsylvania.

Stone, J., and Hoffmann, K. P. (1972). Very slow-conducting ganglion cells in the cat's retina: a major new functional type? *Brain Res.* **43**, 610.

Villars, F. M. H., and Benedek, G. B. (1974). "Physics with Illustrative Examples from Medicine and Biology," Vol. 2, Statistical Physics. Addison Wesley, Reading, Massachusetts.

von Békésy, G. (1967). "Sensory Inhibition." Princeton, Univ. Press, Princeton, New Jersey.

Wald, G. (1933). Vitamin A in the retina. *Nature (London)* **132**, 316.

Wald, G. (1945). Human vision and the spectrum. *Science* **101**, 653.

Wald, G. (1961). The molecular organization of visual systems. *In* "Light and Life" (W. D. McElroy and B. Glass, eds.). Johns Hopkins Press, Baltimore, Maryland.

Wald, G. (1964). The receptors for human color vision. *Science* **145**, 1007.

Wald, G., and Brown, P. K. (1965). Human color vision and color blindness. *Cold Spring Harbor Symp. Quant. Biol.* **30**.

Wald, G., Brown, P. K., and Gibbons, I. R. (1963). The problem of visual excitation. *J. Opt. Soc. Am.* **53**, 20.

Werblin, F. S., and Dowling, J. E. (1969). Organization of the retina of the mudpuppy *Necturus maculosus* II. Intracellular recording. *J. Neurophysiol.* **32**, 339.

Wiesel, T. N., and Hubel, D. H. (1966). Spatial and chromatic interactions in the lateral geniculate body of the rhesus monkey. *J. Neurophysiol.* **29**, 1115.

Wolff, E. (1954). "Anatomy of the Eye and Orbit." Lewis, London.

Wormington, C. M., and Cone, R. A. (1978). Ionic blockage of the light-regulated sodium channels in isolated rod outer segments. *J. Gen. Physiol.* **71**, 657.

Wright, W. D. (1928). A re-determination of the trichromatic coefficients of the spectral colors. *Trans. Opt. Soc. London,* **30**, 141.

Wright, W. D., and Pitt, F. H. G. (1934). *Proc. R. Soc. London. Ser. A* **44**, 459.

Yoshikami, S., and Hagins, W. A. (1971). Light, calcium and the photocurrent of rods and cones. *Biophys. J. Abstr.* **11**, 47a (Fifteenth Annual Meeting of the Biophysical Society, 1971).

CHAPTER 7

Psychophysics

INTRODUCTION

The term *psychophysics*, coined over a hundred years ago, is used in reference to the scientific discipline concerned with the responses that organisms make to energies of the environment. The field is concerned with the relationship between sensation, a subjective value, and external signal, an objective value. The main thrust of this book has been *how* perception is accomplished, and it is appropriate at this point to consider *what* is perceived. Psychophysics tells us what an organism can do and, from this approach, it provides useful information to the electrophysiologist in the interpretation of the electrical impulses which he measures in neurons.

The interpretation of psychophysical measurements has been a matter of extreme controversy during its century-old history. Part of this controversy seems to have been caused by the inability on the part of physicists and psychologists simply to understand each other's viewpoint with the result that measurements have not been made, let alone interpreted, to their mutual satisfaction. The more recent impetus for researchers in the two disciplines to work together toward a mutual understanding seems to have come from the commercial world. For example, the acoustical energy from a loudspeaker must be reduced by 90% for a sound to seem half as loud to a listener regardless of the original level, and energy output of a light must be reduced by the same amount to seem half as bright. However, to make one weight seem half as heavy as another it must be reduced by only 38%. These and other measurements attract both commercial and military interest and considerable funding of research has been supplied by them. The necessary measurements have been made and some agreement achieved, although the reasons why the response of organisms vary with varying stimuli still remain speculative. In this chapter we will briefly summarize the types of investigation and their findings.

SUBJECTIVE MEASUREMENT

The first problem encountered by an investigator of psychophysical laws is the decision of what can be investigated quantitatively. For example, the beautiful–ugly scale to a given individual is very real but, not only is the scale fraught with cultural bias, it does not lend itself to a physical measurement which can be compared with the subjective measure (Weyl, 1952). Thus, abstraction or designs are not measureable. However, components of these can be used: the perceived relative angle or length of lines can be compared to their physical measurements. In this way the resolving power of visual perception of two quantities can be measured as a function of their magnitude. Loudness, pitch, brightness, weight, pressure, taste and olfaction intensity of varying quantities of the same substance are other examples. We have seen in Chapter 5 that perception of changes in two stimuli by the same organ differs, such as loudness and pitch by the ear.

CLASSIFYING STIMULI

Stevens (1957) has proposed that perceptual measurements be divided into two classes which separate quantity from quality. In this division Class I (called prothetic) is concerned with *how much*, for example, loudness. Class II (called metathetic) is concerned with *what kind* or *position*, for example, sound pitch or location. Four functional criteria are relevant to the distinction between these two classes.

(1) Size of the measuring unit. Suppose each unit is taken as the same size. It would then be expected that a sound at constant frequency with an energy of 100 units would seem to the subject to be twice as loud as one of

50 units. But we have already seen that this is not the case. We conclude that in loudness measurement, which is a Class I perception, units are not additive. In contrast, a measurement of a Class II perception, such as sound pitch (frequency), shows that units are essentially additive.

(2) Rating scales. When a subject is asked to judge differences between sensations he may or may not be supplied with a rating scale. For example, he is to judge the relative duration of noise and is told the shortest and longest durations. He may then either be asked to make his own scale of relative lengths or be asked to judge them over a scale of 1–10. When he is given a scale, the results always concave downward when the category rating is plotted against the duration, Fig. 7.1. If the subject makes up his own scale, the result is often linear, Fig. 7.1. The reason for this is the following. If the subject tries to put duration into preset categories of equal intervals, such as 1–7 in Fig. 7.1, he is biased in that a given difference at the low end is more noticeable than at the high end. Since he can easily tell 0.5 sec from 1.0 sec he tends to put them in different categories, but since he can tell only with difficulty the difference between 3.5 and 4 sec he tends to put them in the same category. With Class II stimuli, such differences in curvature are not noted except near the upper and lower thresholds.

(3) Time–order error. When a subject is asked to judge between the magnitude of two Class I stimuli and the experimenter presents him with two equal stimuli, the subject tends to judge the second as greater than the first. This does not happen for Class II stimuli. The time–order error is

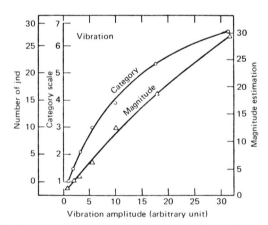

FIG. 7.1 Perceived intensity versus stimulus intensity of a 60 Hz vibration applied to finger tip. △, mean magnitude estimation by 12 subjects. ○, mean judgements by 16 subjects on a scale of 1 to 7. [Reprinted from Stevens (1961) by permission of the MIT Press. Copyright 1961 by the Massachusetts Institute of Technology.]

small, however, and is reversed in organs that fatigue, for example, flashes of light on the retina.

(4) Hysteresis. In a measurment of subjective loudness a different curve is developed if the loudness is successively increased than when it is decreased. It is not a hysteresis effect such as in physical phenomena, but the two curves obtained resemble a hysteresis loop. It has been shown to occur in comparative brightness measurements and in relative magnitudes of lifted weights. Hysteresis occurs in Class I perception but not in Class II.

THE LIMEN

If a scale of sensation is to be established, it must have units. The lower limit is called the *absolute stimulus threshold*. Although it would seem that this should be the point at which a subject can just detect a stimulus, it is more complicated. All measurements are statistical and there are no absolutes. Thus, the absolute threshold or any other measurements must be made many times and an agreed statistical probability must be used. This level is generally taken above 50%, usually 75%. In the case of absolute threshold, for example, it would be taken as that external stimulus which is detectable in 75% of the tests on a given subject. The upper limit, called the *terminal stimulus* or *terminal threshold*, can arise for different reasons: (a) it can be the upper limit of detectability of the organ, such as frequency of sound in hearing or of light in vision, or (b) a saturation of sensitivity of the nervous system beyond which it can no longer discriminate differences.

The units of the above scale are frequency called *just notable differences* or jnd, for example, the frequency difference or loudness difference which is just detectable. This discrimination difference above or below a given reference point is also called a *limen*; a difference greater than a jnd is called *supralimenal* while one below detectability is called *sublimenal*. As before, it should be noted that the size of a limen is statistical and again a 75% detectability is usually taken as the criterion for specification.

WEBER AND FECHNER LAWS

If a subject is to judge comparatively two weights, their weight difference is not the important parameter. For example, he can readily distinguish between weights of 1 and 2 gm. but not between 1000 and 1001 gm. Similarly, he can distinguish between lines of length 1 and 2 cm but not between 100 and 101 cm, or the brightness change between one and two candles but not between 1000 and 1001 candles.

In the last century Weber recognized the above observations and stated "in comparing magnitudes, it is not the arithmetical difference, but the ratio of the magnitudes which we perceive" (Thurstone, 1948). This is known as Weber's law. Although this statement is clear, it is not in an operational form, that is, it does not explicitly define a laboratory experiment. Without loss of generality, the law is usually reworded as "the difference limen is a constant fraction of the stimulus", and in practice the magnitude of the stimulus is taken as the average of the two stimuli which can just barely be perceived. In this form it may be written mathematically as

$$dI/I = k \qquad (7.1)$$

where I is the intensity of the stimulus from which the limen dI is measured and k is a constant. The constant k is usually not the same for different types of stimuli.

The formulation of Weber's law is still not adequate for psychophysical measurements because dI is a single limen. Suppose two quite different weights are given to a subject as well as a varied assortment of weights from which he is to choose one half way between the two. It is found experimentally that the subject does not choose the correct one and it is these differences in perception in relation to reality which are to be measured. To consider this problem, the incremental limen dI must be replaced by a ΔI but, if I represents the actual weight, what represents the perceived weight?

Following Weber, in the last century Fechner introduced a perception factor S and wrote the equation

$$dS = k \, dI/I \qquad (7.2)$$

where dS is a perception increment, or limen, and k, while still a constant, is a different one from that in Eq. (7.1). The integral of Eq. (7.2) is

$$S = k \log I + \text{const} \qquad (7.3)$$

and this relation is known as Fechner's law. Thus, while Weber's law is concerned with limenal (or sublimenal) differences in which the subject has some difficulty in discrimination, Fechner's law deals with supralimenal differences in which the fraction of correct discriminations is unity. Under conditions of small differences, the two laws can be satisfied by the same set of experimental data, but under conditions of large differences one of these laws can be satisfied while the other fails.

The validity of Fechner's law has been debated for a century. Fechner assumed that the magnitudes of the limen are constant and that, therefore, the magnitude of perception differences are constant. Since dI/I in physics terminology is the resolving power, Fechner's assumption is equivalent to saying that the measures of resolving power provide equal units that can be used as a measure of magnitude. This assumption is incorrect, as will be shown below (Stevens, 1957).

THE POWER LAW

When the perception S, or psychological magnitude, is measured against a physical magnitude I, the resulting data depend upon the type of stimulus. Three examples of this for brightness of illumination, apparent line length, and electric shock intensity are shown in Fig. 7.2. Note that these are Class I stimuli. If, instead of a plot on linear coordinates, the graph is made on log–log coordinates as in Fig. 7.3, straight lines result, although with different slopes.

Recalling from Eq. (7.2) that S in our notation is perception (or psychological) magnitude and I is stimulus magnitude, the equation of the straight lines in Fig. 7.3 may be written as

$$\log S = n \log I + \log K \qquad (7.4)$$

where n is the slope and $\log K$ the intercept. The antilog of Eq. (7.4) is

$$S = KI^n \qquad (7.5)$$

which is known as the *power law* of psychological perception. A more common form of Eq. (7.5) is

$$S = K(I - I_0)^n \qquad (7.6)$$

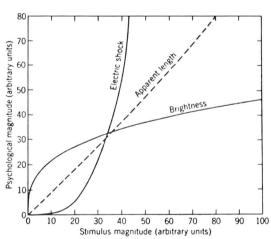

FIG. 7.2 Linear plot of psychological magnitude versus stimulus for three different types of measurements. [Reprinted from Stevens (1961) by permission of the MIT Press. Copyright 1961 by the Massachusetts Institute of Technology.]

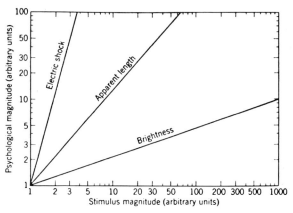

FIG. 7.3 The data of Figure 7.2 plotted on log–log scale. [Reprinted from Stevens (1961) by permission of the MIT Press. Copyright 1961 by the Massachusetts Institute of Technology.]

where I_0 is a small constant corresponding to the absolute stimulus threshold.

Figure 2.11 illustrates the applicability of the power law to taste and shows that the neural response, frequency of action potentials along the chorda typani, has the same slope, i.e., exponent n, as the psychological sensation. The shift of the intercept occurs because no normalization of

TABLE 7.1

Exponents of the Power Function (Eq. (7.5)) Relating Psychological magnitude to Stimulus Magnitude for Various Continua[a]

Continuum	Exponent	Stimulus condition
Loudness	0.6	Biaural
Loudness	0.54	Monaural
Brightness	0.5	Point source
Brightness	0.33	5° target
Smell	0.55	Coffee
Smell	0.6	Heptane
Taste	1.3	Sucrose
Taste	1.3	Salt
Temperature	1.0	Cold on arm
Temperature	1.6	Warmth on arm
Vibration	0.95	60 Hz on finger
Duration	1.1	White noise stimulus
Pressure	1.1	Static force on palm
Heaviness	1.45	Lifted weight
Force of handgrip	1.7	Precision hand dynamometer
Electric shock	3.5	60 Hz through fingers

[a] From Stevens (1961).

psychological to electrophysiological measure was made. Figure 7.4 shows that response R of single mechanoreceptive nerve fibers in a cat to variations of pressure also follows a power law with a slope similar to that obtained by psychophysical measurements on humans.

As seen in Fig. 7.3, different stimuli give rise to different slopes, or exponents in the power law. Table 7.1 gives average exponents for some selected stimuli. It must be kept in mind that not only are the values in Table 7.1 averages of experiments, subjects, and different laboratories, but each measurement is subject to the Class I difficulties discussed earlier.

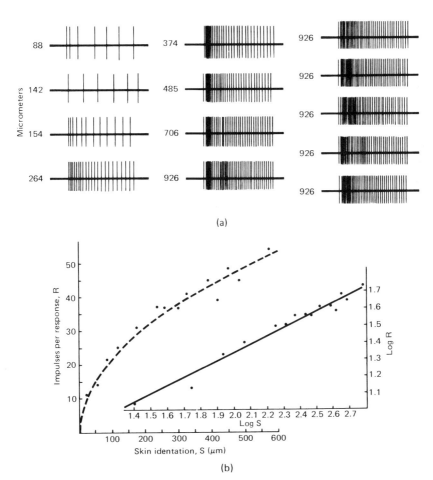

FIG. 7.4 (a) Nerve impulses evoked in a single fiber of the saphenous nerve (in the thigh) of a cat by mechanical indentation of its ending in the skin of the lower leg. (b) A linear and a log–log plot of the response R versus skin indentation S in microns. The data obey a power law $R = KS^n$ with $K = 2.16$ and $n = 0.509$. [From Werner and Mountcastle (1965).]

CROSS-MODALITY COMPARISONS

Physical scientists often have skepticism about measurements which involve human judgement with the associated statistical distribution. Some of this skepticism may be allayed by consideration of cross-modality measurements. For example, can a subject give a reasonable opinion between the equality of intensity of vibration on his finger with sound in his ears? Such experiments have been done. If v is vibration and l is loudness, the simple power law, Eq. (7.5), would be for each, with an appropriate choice of units.

$$S_v = I_v^n, \qquad S_l = I_l^m \tag{7.7}$$

If the subject equates the two sensations at seemingly equal intensities, then $S_v = S_l$ and

$$I_v^n = I_l^m$$

or

$$\log I_v = (m/n)\log I_l \tag{7.8}$$

Experimental results of such measurements are illustrated in Fig. 7.5 where the abscissa and ordinate are logarithmic, i.e., dB = decibel. The squares represent results when vibration is adjusted to match noise and the circles represent the converse. The slope of this line, log vibration/log noise is 0.6, which is reasonably close to that predicted by Table 7.1 of 0.6/0.95.

Another example of cross-modality measurement is that of matching the

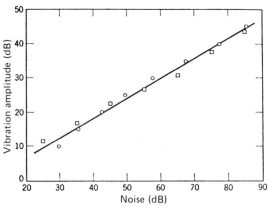

FIG. 7.5 Graph of equal sensation between a 60-Hz vibration on the finger and the sound intensity of a band of noise. O, the subjects adjusted loudness to match vibration. □, the subjects adjusted vibration to match loudness. Decibels (dB) are logarithmic quantities. [From Stevens (1959b). Copyright 1959 by the American Psychological Association. Reprinted by permission.]

relative brightness of two 5 deg spots of light to the relative loudness of two sounds. First, the power laws of loudness and brightness were obtained. These are shown in Fig. 7.6. Note that the exponents are approximately 0.3 for both loudness and brightness. In this case the loudness is expressed in terms of sound energy, while the 0.6 value in Table 7.1 is for sound pressure. As shown in Vol. I, Chap. 8, energy is the square of pressure. Thus, in common terms of energy both loudness and brightness have the same slope. The experiment on the cross-modality of the two different stimuli produced the results shown in Fig. 7.7. It is seen that a 45 deg line results which shows that the two exponents are the same, as indicated in Fig. 7.6. These results, as pointed out by Stevens, suggest that the natural unit of light ratios should be decibels as is the convention in sound intensity measurements.

A variety of cross-modalities have been measured. The results of an impressive series of experiments are shown in Table 7.2. In this table the relative intensities of a variety of stimuli are shown versus the equal sensation of a handgrip of varying strengths. It is seen that the results of the experimental ratios compare very favorably to the predicted exponent, obtained from the separate exponents of Table 7.1. See Anderson (1979) for a discussion of the algebra of different modalities.

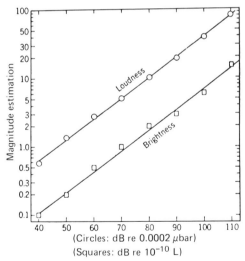

FIG. 7.6 Log–log plots of perception of loudness and brightness versus stimulus intensity. For loudness a 1000 Hz tone was used. Brightness was measured by a 3-sec illumination of a target which subtended a 5-deg angle from the dark-adapted eye. [From Stevens (1957). Copyright 1957 by the American Psychological Association. Reprinted by permission.]

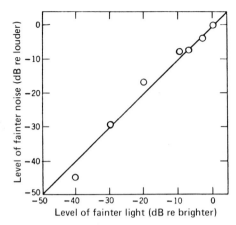

FIG. 7.7 Results of adjusting loudness ratio to match the apparent brightness ratio of a pair of luminous circles. One of the circles was made dimmer than the other by the amount shown on the abscissa. The observer adjusted the level of one noise (the ordinate) to make the loudness ratio seem equal to the brightness ratio. [Reprinted from Stevens (1961) by permission of the MIT Press. Copyright 1961 by the Massachusetts Institute of Technology.]

TABLE 7.2

The Exponents (Slopes) of Equal-Sensation Functions, as Predicted from Ratio Scales of Subjective Magnitude, and as Obtained by Matching with Force of Handgrip[a]

Ratio scale		Scaling by means of handgrip		
Continuum	Exponent of power function	Stimulus range	Preducted exponent	Obtained exponent
Electric shock (60-Hz current)	3.5	0.29–0.72 mA	2.06	2.13
Temperature (warm)	1.6	2.0–14.5°C above neutral temperature	0.94	0.96
Heaviness of lifted weights	1.45	28–480 gm	0.85	0.79
Pressure on palm	1.1	0.5–5.0 lbs	0.65	0.67
Temperature (cold)	1.0	3.3–30.6°C below neutral temperature	0.59	0.60
60-Hz vibration	0.95	17–47 dB reapproximate threshold	0.56	0.56
Loudness of white noise	0.6	55–95 dB re 0.0002 dyn/cm^2	0.35	0.41
Loudness of 1000-Hz tone	0.6	47–87 dB re 0.0002 dyn/cm^2	0.35	0.35
Brightness of white light	0.33	56–96 dB re 10^{-10}	0.20	0.21

[a] From Stevens (1961), based on data from Stevens *et al.* (1960) and Stevens and Stevens (1960).

INTERPRETATION OF THE POWER LAW

The examples of the applicability of the power law relating preceived sensation to external stimuli represent the briefest of summaries of an overwhelming amount of evidence for its validity. Why is the organism constructed this way? Clearly, the transduction of external information into preception must compress or expand the scale for the processing of stimuli required for survival. That much is obvious and therefore it is not expected that the same sensitivity ratios prevail for the various type of receptors. But why a logarithmic behavior? We will follow the arguments of Rushton (1961) in suggesting the reason.

Very simple animals and, presumably early forms of life, have no nerves. They do react to stimuli, however, by the secretion of chemicals in response to an external stimulus, which in turn causes some activity on their part. For example, they may move their flagellae to swim up a food gradient or away from a noxious gradient. With larger life forms, the organism cannot wait for the secretion of a chemical from some organ to diffuse by way of the bloodstream. In response to certain external stimuli, it must react almost immediately or its species would not have survived. Evolution developed the nerves to signal rapidly the excretion of chemicals from thousands of different nerve endings rather than wait for a hormonal response by diffusion alone. Thus, nerves can be thought of as a rapid signal device to distant chemical transmitters.

We have seen in Fig. 6.45 that the change in membrance resistance of the eye of the *Limulus* is proportional to $\log(I_0 + I)$, where I is the light intensity and I_0 represents a fixed level of "noise" in the system, even in the dark, which is equivalent to a low level illumination. The steady depolarization of the cell apparently arises from the arrival of impulses along the nerve at frequency $n/\sec$. The fact that the frequency of impulses and the membrane potential of Fig. 6.44 are both uniform scales means that they are linearly related to each other and hence to the change in membrane resistance. The fact that the change in membrane resistance is caused by the liberation of some transmitted chemical means that the rate of production of this chemical must be proportional to $\log(I_0 + I)$. Although this is an elementary model, it is worth pursuing to a conclusion.

Suppose that the transmitter chemical C_1 has a concentration proportional to $\log(I_0 + I)$ and that this chemical causes a drop in the membrane resistance and potential and a proportional rise in the frequency of impulses n_1 generated in the nerve. At the other end of the nerve some other (or similar) chemical C_2 is released in packets, one for each of the impulses n_1. If C_2 is continuously removed, its steady-state concentration will be proportional to n_1 or $C_2 = a_1 n_1$, where a_1 is a constant. This could proceed linearly down a chain of synapses, but organisms are not con-

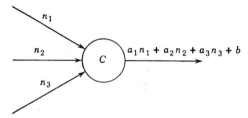

FIG. 7.8 Schematic of the impulse rates n from three nerves converging on a synapse C and the resulting output. Any of the n's may be negative if the impulse is inhibitory. [Reprinted from Rushton (1961) by permission of the MIT Press. Copyright 1961 by the Massachusetts Institute of Technology.]

structed in this manner. That is, a single nerve fiber continues until it synapses with other nerves, not just a continuation of itself. Thus, a diagram as in Fig. 7.8 would be more realistic in which several nerve fibers converge upon a single outgoing fiber and the outgoing impulses would be a weighted mean of the incoming ones, as indicated. Some of the incoming impulses may be inhibitory which would simply give a negative value to the corresponding a coefficients. The experiments of Hartline and Ratliff (1957) and Fuortes (1959) demonstrate this relation.

Sense organs in highly developed animals must cover a wide range of stimuli, of about 50–100 dB. At the lower level of intensity, the signals are so weak that they approach the noise level of the sensing organ while at high intensity the requirement is not merely to detect but to discriminate with respect to location, movement, pitch, etc. If the required jnd for discrimination at the low stimulus intensities were a constant throughout the range of intensity, then the organism could not detect small intensity changes at high intensities. Conversely, a jnd at high intensities would overwhelm a sense organ at low intensities. Therefore, the resolving power should vary with the intensity. That is, the ratio of just noticeable difference ΔI to the intensity I should produce a constant output Δn to the organism or $\Delta I / I = \Delta n$. We have seen that this is Weber's law, Eq. (7.1).

The eyes, ears, and nose are long distance receptors and the organism need not know absolute intensities but only changes, differences, or patterns. As illumination changes, for instance, the patterns of objects change in the same ratio. If I is the mean intensity and $I + \Delta I_p$ the intensity at point p, the pattern of $\Delta I_p / I$ will be the same at all intensity levels. This invariance of $\Delta I_p / I$ is the same relation as that of Weber's law. Thus, there is a dual significance of this relation and of the logarithmic relation between the intensity of stimulus I and the frequency of impulses n. This is readily seen if we write the relation that the frequency impulses n

above some n_0 is proportional to the logarithm of intensity

$$a(n - n_0) = \log I \tag{7.9}$$

and, writing this as the variation, gives

$$a\,\Delta n = \Delta I / I \tag{7.10}$$

As seen earlier in this section, $a\Delta n$ is the signal transmitted through the axon without change. As the light intensity changes the discrimination of the pattern remains unchanged.

REFERENCES

Anderson, N. (1979). Algebraic rules in psychological measurements. *Am. Sci.* **64**, 555.

Fuortes, M.G.F. (1959). Initiation of impulses in the visual cells of *Limulus*. *J. Gen. Physiol.* **148**, 14.

Hartline, H.K., and Ratliff, F. (1957). Inhibiting interaction of receptor units in the eye of *Limulus*. J. Gen. Physiol. **40**, 157.

Rushton, W. A. H. (1961). Peripheral coding in the nervous system. *In* "Sensory Communication," (W. A. Rosenblith, ed.). MIT Press, Cambridge, Massachusetts.

Stevens, S. S. (1957). On the psychophysical law. *Psychol. Rev.* **64**, 153.

Stevens, S. S. (1959a). Tactile vibration: dynamics of sensory intensity. *J. Exp. Psychol.* **57**, 210.

Stevens, S. S. (1959b). Cross-modality validation of subjective scales for loudness, vibration and electric shock. *J. Exp. Psychol.* **57**, 201.

Stevens, S. S. (1961). The psychophysics of sensory function. *In* "Sensory Communication" (W. A. Rosenblith, ed.). MIT Press, Cambridge, Massachusetts.

Stevens, J. C., and Stevens, S. S. (1960). Warmth and cold: dynamics of sensory intensity. *J. Exp. Psychol.* **60**, 183.

Stevens, J. C., Mack, J. D., and Stevens, S. S. (1960). Growth sensation on seven continua as measured by the force of handgrip. *J. Exp. Psychol.* **59**, 60.

Thurstone, L. L. (1948). Psychophysical methods. *In* "Methods of Psychology" (T. G. Andrews, ed.). Wiley, New York, 1948.

Werner, G., and Mountcastle, V. B. (1965). Neural activity in mechano-receptive cutaneous afferents: stimulus-response relations, Weber functions and information transmission. *J. Neurophysiol.* **28**, 359.

Weyl, H. (1952). "Symmetry." Princeton Univ. Press, Princeton, New Jersey.

Root Mean Square Deviation

If one measures the number of events of some fluctuating phenomenon in a lengthy time interval t, there will be an average number $\bar{n}_t$ per unit time. If, however, one takes a small sample of events n in a short time, the long time average is generally not obtained; deviations of both greater and less than the average will be observed. The average of such deviations is $\overline{(n - \bar{n}_t)} = 0$ because there are equal numbers of deviations of both plus and minus the average. Therefore, the above simple average does not yield the average deviation. If, however, this quantity is squared, then both plus and minus deviations are positive and the square root of this squared average will give a measure of the mean deviation. This is called the root mean square (rms),

$$\text{rms deviation} = \left[\overline{(n - \bar{n}_t)^2} \right]^{1/2} \tag{A.1}$$

It should be noted here that the difference $(n - \bar{n}_t)$ is sometimes referred to as the fluctuation amplitude and written as Δn.

Eq. (A.1) can be expressed in terms of $\bar{n}_t$ alone, or

$$(\bar{n}_t)^{1/2} = \left[\overline{(n - \bar{n}_t)^2} \right]^{1/2}$$

and we will now derive this relation.

First, complete the square of the mean squared deviation

$$\overline{(n - \bar{n}_t)^2} = \overline{n^2 - 2n\bar{n}_t + \bar{n}_t^2}$$
$$= \overline{n^2} - 2\bar{n}\bar{n}_t + \bar{n}_t^2$$

For reasonable values of n, the middle term on the right-hand side can be written as $2\bar{n}_t^2$, and therefore

$$\overline{(n - \bar{n}_t)^2} = \overline{n^2} - 2\bar{n}_t^2 + \bar{n}_t^2$$
$$= \overline{n^2} - \bar{n}_t^2 \qquad (A.2)$$

We now need to obtain a value for $\overline{n^2}$ from our knowledge of the Poisson distribution.

Let us change the concept of Eq. (A.1) in the following way. We wish to find the probability of n events occurring when $\bar{n}_t$ is the average number of events. Equation (6.1) is then

$$P_n = \frac{\bar{n}_t^n e^{-\bar{n}_t}}{n!} \qquad (A.3)$$

By the rules of calculus, an average value of a quantity is obtained by integrating (or summing) the product of the function and quantity whose average is to be obtained over all possible values. Thus,

$$\overline{n^2} = \sum_{n=0}^{\infty} n^2 P_n = \sum_{n=0}^{\infty} n^2 \frac{\bar{n}_t^n e^{-\bar{n}_t}}{n!}$$
$$= e^{-\bar{n}_t} \sum_{n=0}^{\infty} \frac{n^2 \bar{n}_t^n}{n!}$$

Writing some terms of this series yields

$$\overline{n^2} = e^{-\bar{n}_t} \left[\bar{n}_t + \frac{2^2}{2!} \bar{n}_t^2 + \frac{3^3}{3!} \bar{n}_t^2 + \frac{4^4}{4!} \bar{n}_t^2 + \cdots \right]$$
$$= \bar{n}_t e^{-\bar{n}_t} \left[1 + 2\bar{n}_t + \frac{3}{2!} \bar{n}_t^2 + \frac{4}{3!} \bar{n}_t^3 + \cdots \right]$$

Recall that the series expansion for an exponential is

$$e^x = 1 + x + \frac{x^2}{2!} + \frac{x^3}{3!} + \cdots$$

The series above in the brackets is the sum of an exponential and the product of an exponential and $\bar{n}_t$. This is seen by rewriting the bracket terms as

$$\overline{n^2} = \bar{n}_t e^{-\bar{n}_t} \left\{ \left[1 + \bar{n}_t + \frac{\bar{n}_t^2}{2!} + \frac{\bar{n}_t^3}{3!} + \cdots \right] + \left[\bar{n}_t + \frac{2}{2!} \bar{n}_t^2 + \frac{3}{3!} \bar{n}_t^3 + \cdots \right] \right\}$$

$$= \bar{n}_t e^{-\bar{n}_t} \left\{ e^{\bar{n}_t} + \bar{n}_t e^{\bar{n}_t} \right\}$$

$$= \bar{n}_t + \bar{n}_t^2 \qquad\qquad (A.4)$$

Upon substituting Eq. (A.4) into Eq. (A.2), we obtain

$$\overline{(n - \bar{n}_t)^2} = \bar{n}_t + \bar{n}_t^2 - \bar{n}_t^2$$

or

$$\overline{(n - \bar{n}_t)^2} = \bar{n}_t$$

and

$$\left[\overline{(n - \bar{n}_t)^2} \right]^{1/2} = (\bar{n}_t)^{1/2} \qquad\qquad (A.5)$$

Thus, the rms deviation varies as the square root of the average of the total number of events. If we define a signal S as the average of a large number of events $\bar{n}_t$ and the noise as the root mean squared deviation from the average $[\overline{(n - \bar{n}_t)^2}]^{1/2} = (\bar{n}_t)^{1/2}$, then the signal-to-noise ratio K is

$$K = S/\mathfrak{N} = \bar{n}_t / (\bar{n}_t)^{1/2} = (\bar{n}_t)^{1/2} \qquad\qquad (A.6)$$

Index

271